DELMAR

THOMSON LEARNING

Fiber Optics Technician's Manual, Second Edition
by Jim Hayes and the Instructors at Fiber U

Business Unit Director:
Alar Elken

Executive Editor:
Sandy Clark

Acquisitions Editor:
Mark W. Huth

Development:
Dawn Daugherty

Executive Production Manager:
Mary Ellen Black

Project Editor:
Ruth Fisher

Production Coordinator:
Ruth Fisher

Art/Design Coordinator:
Rachel Baker

Executive Marketing Manager:
Maura Theriault

Channel Manager:
Mona Caron

Marketing Manager:
Kasey Young

For permission to use material from this text or product, contact us by
Tel 1-800-730-2214
Fax 1-800-730-2215
www.thomsonrights.com

Library of Congress Cataloging-in-Publication Data

Hayes, Jim, 1946–
    Fiber optics technician's manual / Jim Hayes and the instructors of Fiber U.—2nd ed.
        p. cm.
    ISBN 0-7668-1825-X (core text : pbk. : alk. paper) —
ISBN 0-7668-1826-8 (IG : pbk. : alk. paper)
    1. Optical communications—Handbooks, manuals, etc.
2. Fiber optics—Handbooks, manuals, etc. 3. Telecommunication cables—Handbook, manuals, etc. I. Title.
TK5103.59. H39 2001
621.36′92—dc21   00-025836

NOTICE TO THE READER

FIBER OPTICS TECHNICIAN'S MANUAL

D1361900

FIBER OPTICS TECHNICIAN'S MANUAL

2nd Edition

Jim Hayes
and the Instructors of Fiber U

DELMAR
™
THOMSON LEARNING

Australia Canada Mexico Singapore Spain United Kingdom United States

CONTENTS

INTRODUCTION

Fiber optics has become the transmission medium of choice for most communications. Its high speed and long-distance capability make it the most cost-effective communications medium for most applications. While higher performance and lower cost components are in continual development, it has also become critical to train competent personnel to design, install, and maintain state-of-the-art fiber optic networks.

This book is intended as a textbook for training installers of fiber optic networks. It includes only an introduction to the basics of fiber optics, leaving the physics and mathematics of fiber optic technology to high-level textbooks. Instead, this book focuses on the practical aspects of designing, installing, testing, and troubleshooting fiber optic cable plants and networks.

This book has been edited from the training programs of the Fiber U fiber optic training conferences and those developed by a number of professional instructors in fiber optics. The Fiber U conference was developed as an annual training program that offered a combination of classroom seminars and multi-vendor hands-on training. Fiber U training also kept pace with the changes in technology, with each year's presentations and products being the latest.

A unique characteristic of Fiber U is its practical orientation to the installer and user of the fiber optic cable plant. It provided a basic introduction to fiber optic technology, but skipped the complex background and theory included in most textbooks and courses. By concentrating on understanding the basics of fiber optics and the practical aspects of designing, installing, testing, and troubleshooting the cable plant, it provided an intensive learning experience that translated into practical applications immediately.

The Fiber U program was developed by a number of instructors, all of whom are involved in teaching fiber optics courses regularly. The material comes from their practical experience and the assistance of a large number of vendors who provided the latest product information. Those instructors have joined together to produce this book.

The timeliness of the material is important in a technology such as fiber optics that is moving rapidly. New product innovation and rapid cost reduction are the norm for fiber optics. We have tried to include the latest material as of our publishing date. While we intend to update the material regularly, it is always a good idea to contact the vendors of fiber optic products for the latest applications and product information.

This book includes an interesting history of the development of fiber optics and an overview of the basics of the technology to give perspective to the reader. Then the material is oriented to describing the components of a fiber optic network and how to choose, install, test, and troubleshoot them. A glossary of terms and resource guide are included for reference material.

An industry-recognized certification program has developed by the Fiber Optic Association from Fiber U, which uses much information contained in this book. This book may be effectively used as a study guide for the FOA CFOT certification test.

The authors welcome your feedback on how we can improve this book and make it more useful to you.

A NOTE OF APPRECIATION

We wish to thank all those who contributed material and support to Fiber U and this book by providing us with product and technical information for inclusion. We also especially wish to thank Hugh Doherty of 3M who, during a coffee break at an FDDI conference in Sunnyvale, CA, in April of 1992, convinced us (well, me, Jim Hayes of Fotec) to undertake the massive project that became Fiber U; Debra Norman of *Fiberoptic Product News* who suggested the name "Fiber U"; and Wayne Kachmar of Northern Lights Cable who showed us how to use a little showmanship to make a big impression on trainees.

So our thanks to all these companies and people who made it possible, including the companies sponsoring the Fiber U conferences and especially our coauthors who have contributed their time and effort to this book:

Professor Elias Awad
Dave Chaney
Trevor Conquest, Conquest Communications, Australia
Thomas A. Dooley, Fiber Specialists, Inc.
Mark Duquesne
Douglas F. Elliot, IBEW, Canada
Jim Hartman, FiberLite

Jeff Hecht
John Highhouse, Lincoln Trail Community College
Larry Johnson, The Light Brigade
Michael Kovacs, Kovacs Engineering
Steve Paulov, Cabling Business
Eric Pearson, President, Pearson Technologies
Paul Rosenberg
Professor Jerald R. Rounds, Arizona State University
Phil Sheckler, Technology Standards Group

A special thanks from Delmar and the authors for the suggestions from the reviews of this edition:

Clarence Swallow, Northwest Kansas Technical School
Robert Edgecomb, Aviation and Electronic School of America

THE ORIGINS OF FIBER OPTIC COMMUNICATIONS

JEFF HECHT

Optical communication systems date back two centuries, to the "optical telegraph" invented by French engineer Claude Chappe in the 1790s. His system was a series of semaphores mounted on towers, where human operators relayed messages from one tower to the next. It beat hand-carried messages hands down, but by the mid-19th century it was replaced by the electric telegraph, leaving a scattering of "telegraph hills" as its most visible legacy.

Alexander Graham Bell patented an optical telephone system, which he called the Photophone, in 1880, but his earlier invention, the telephone, proved far more practical. He dreamed of sending signals through the air, but the atmosphere did not transmit light as reliably as wires carried electricity. In the decades that followed, light was used for a few special applications, such as signaling between ships, but otherwise optical communications, such as the experimental Photophone Bell donated to the Smithsonian Institution, languished on the shelf.

Thanks to the Alfred P. Sloan Foundation for research support. This is a much expanded version of an article originally published in the November 1994 *Laser Focus World*.

In the intervening years, a new technology that would ultimately solve the problem of optical transmission slowly took root, although it was a long time before it was adapted for communications. This technology depended on the phenomenon of total internal reflection, which can confine light in a material surrounded by other materials with lower refractive index, such as glass in air.

In the 1840s, Swiss physicist Daniel Collodon and French physicist Jacques Babinet showed that light could be guided along jets of water for fountain displays. British physicist John Tyndall popularized light guiding in a demonstration he first used in 1854, guiding light in a jet of water flowing from a tank. By the turn of the century, inventors realized that bent quartz rods could carry light and patented them as dental illuminators. By the 1940s, many doctors used illuminated Plexiglas tongue depressors.

Optical fibers went a step further. They are essentially transparent rods of glass or plastic stretched to be long and flexible. During the 1920s, John Logie Baird in England and Clarence W. Hansell in the United States patented the idea of using arrays of hollow pipes or transparent rods to transmit images for television or facsimile systems. However, the first person known to have demonstrated image transmission through a bundle of optical fibers was Heinrich Lamm (Figure 1-1), then a medical student in Munich. His goal was to look inside inaccessi-

Figure 1-1 Heinrich Lamm as a German medical student in 1929, about the time he made the first bundle of fibers to transmit an image. Courtesy Michael Lamm

Figure 1-2   Holger Møller Hansen in his workshop.
Courtesy Holger Møller Hansen

ble parts of the body, and in a 1930 paper he reported transmitting the image of a light bulb filament through a short bundle. However, the unclad fibers transmitted images poorly, and the rise of the Nazis forced Lamm, a Jew, to move to America and abandon his dreams of becoming a professor of medicine.

In 1951, Holger Møller Hansen (Figure 1-2) applied for a Danish patent on fiber optic imaging. However, the Danish patent office denied his application, citing the Baird and Hansell patents, and Møller Hansen was unable to interest companies in his invention. Nothing more was reported on fiber bundles until 1954, when Abraham van Heel (Figure 1-3), of the Technical University of Delft

Figure 1-3   Abraham C. S. van Heel, who made clad fibers at the Technical University of Delft. Courtesy H. J. Frankena, Faculty of Applied Physics, Technical University of Delft

Figure 1-4   Harold H. Hopkins looks into an optical instrument that he designed. Courtesy Kelvin P. Hopkins

in Holland, and Harold H. Hopkins (Figure 1-4) and Narinder Kapany, of Imperial College in London, separately announced imaging bundles in the prestigious British journal *Nature*.

Neither van Heel nor Hopkins and Kapany made bundles that could carry light far, but their reports began the fiber optics revolution. The crucial innovation was made by van Heel, stimulated by a conversation with the American optical physicist Brian O'Brien (Figure 1-5). All earlier fibers were bare, with total internal reflection at a glass-air interface. Van Heel covered a bare fiber of glass or plastic with a transparent cladding of lower refractive index. This protected the total-reflection surface from contamination and greatly reduced crosstalk between fibers. The next key step was development of glass-clad fibers by Lawrence Curtiss (Figure 1-6), then an undergraduate at the University of Michigan working part-time on a project with physician Basil Hirschowitz (Figure 1-7) and physicist C. Wilbur Peters to develop an endoscope to examine the inside of the stomach (Figure 1-8). Will Hicks, then working at the American Optical Co.,

Figure 1-5 Brian O'Brien, who suggested that cladding would guide light along fiber. Courtesy Brian O'Brien, Jr.

made glass-clad fibers at about the same time, but his group lost a bitterly contested patent battle. By 1960, glass-clad fibers had attenuation of about one decibel per meter, fine for medical imaging, but much too high for communications.

Meanwhile, telecommunications engineers were seeking more transmission bandwidth. Radio and microwave frequencies were in heavy use, so engineers looked to higher frequencies to carry the increased loads they expected with the growth of television and telephone traffic. Telephone companies thought video telephones lurked just around the corner and would escalate bandwidth demands even further. On the cutting edge of communications research were millimeter-wave systems, in which hollow pipes served as waveguides to circumvent poor atmospheric transmission at tens of gigahertz, where wavelengths were in the millimeter range.

Even higher optical frequencies seemed a logical next step in 1958 to Alec Reeves, the forward-looking engineer at Britain's Standard Telecommunications Laboratories, who invented digital pulse-code modulation before World War II. Other people climbed on the optical communications bandwagon when the laser

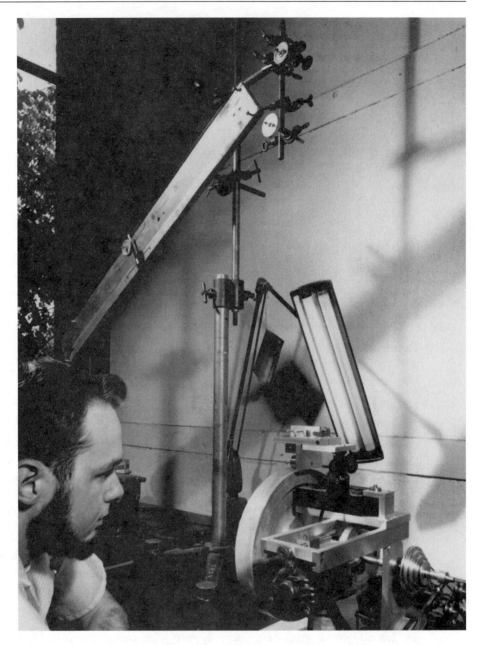

Figure 1-6   Lawrence Curtiss, with the equipment he used to make glass-clad fibers at the University of Michigan. Courtesy University of Michigan News and Information Services Records, Bentley Historical Library, University of Michigan

Figure 1-7   Basil Hirschowitz about the time he helped to develop the first fiber optic endoscope. Courtesy Basil Hirschowitz

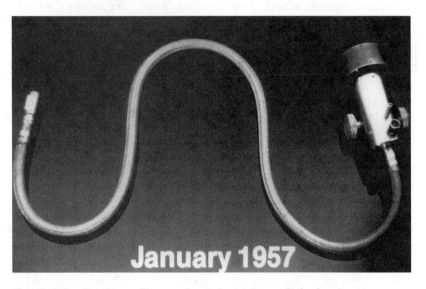

Figure 1-8   Prototype fiber optic endoscope made by Lawrence Curtiss, Wilbur Peters, and Basil Hirschowitz at the University of Michigan. Courtesy Basil Hirschowitz

was invented in 1960. The July 22, 1960, issue of *Electronics* introduced its report on Theodore Maiman's demonstration of the first laser by saying, "Usable communications channels in the electromagnetic spectrum may be extended by development of an experimental optical-frequency amplifier."

Serious work on optical communications had to wait for the CW helium-neon laser. While air is far more transparent to light at optical wavelengths than to millimeter waves, researchers soon found that rain, haze, clouds, and atmospheric turbulence limited the reliability of long-distance atmospheric laser links. By 1965, it was clear that major technical barriers remained for both millimeter-wave and laser telecommunications. Millimeter waveguides had low loss, although only if they were kept precisely straight; developers thought the biggest problem was the lack of adequate repeaters. Optical waveguides were proving to be a problem. Stewart Miller's group at Bell Telephone Laboratories was working on a system of gas lenses to focus laser beams along hollow waveguides for long-distance telecommunications. However, most of the telecommunications industry thought the future belonged to millimeter waveguides.

Optical fibers had attracted some attention because they were analogous in theory to plastic dielectric waveguides used in certain microwave applications. In 1961, Elias Snitzer at American Optical, working with Hicks at Mosaic Fabrications (now Galileo Electro-Optics), demonstrated the similarity by drawing fibers with cores so small they carried light in only one waveguide mode. However, virtually everyone considered fibers too lossy for communications; attenuation of a decibel per meter was fine for looking inside the body, but communications operated over much longer distances and required loss of no more than 10 or 20 decibels per kilometer.

One small group did not dismiss fibers so easily—a team at Standard Telecommunications Laboratories (STL), initially headed by Antoni E. Karbowiak, that worked under Reeves to study optical waveguides for communications. Karbowiak soon was joined by a young engineer born in Shanghai, Charles K. Kao (Figure 1-9).

Kao took a long, hard look at fiber attenuation. He collected samples from fiber makers, and carefully investigated the properties of bulk glasses. His research convinced him that the high losses of early fibers were due to impurities, not to silica glass itself. In the midst of this research, in December 1964, Karbowiak left STL to become chair of electrical engineering at the University of New South Wales in Australia, and Kao succeeded him as manager of optical communications research. With George Hockham (Figure 1-10), another young STL engineer who specialized in antenna theory, Kao worked out a proposal for long-distance communications over singlemode fibers. Convinced that fiber loss should be reducible below 20 decibels per kilometer, they presented a paper at a London meeting of the Institution of Electrical Engineers (IEE). The April 1, 1966, issue of *Laser Focus* noted Kao's proposal:

Figure 1-9 Charles K. Kao making optical measurements at Standard Telecommunications Laboratories. Courtesy BNR Europe

At the IEE meeting in London last month, Dr. C. K. Kao observed that short-distance runs have shown that the experimental optical waveguide developed by Standard Telecommunications Laboratories has an information-carrying capacity . . . of one gigacycle, or equivalent to about 200 tv channels or more than 200,000 telephone channels. He described STL's device as consisting of a glass core about three or four microns in diameter, clad with a coaxial layer of another glass having a refractive index about one percent smaller than that of the core. Total diameter of the waveguide is between 300 and 400 microns. Surface optical waves are propagated along the interface between the two types of glass.

According to Dr. Kao, the fiber is relatively strong and can be easily supported. Also, the guidance surface is protected from external influences. . . . the waveguide has a mechanical bending radius low enough to

Figure 1-10   George Hockham with the metal waveguides he made to model waveguide transmission in fibers. Courtesy BNR Europe

make the fiber almost completely flexible. Despite the fact that the best readily available low-loss material has a loss of about 1000 dB/km, STL believes that materials having losses of only tens of decibels per kilometer will eventually be developed.

Kao and Hockham's detailed analysis was published in the July 1966, *Proceedings of the Institution of Electrical Engineers*. Their daring forecast that fiber loss could be reduced below 20 dB/km attracted the interest of the British Post Office, which then operated the British telephone network. F.F. Roberts, an engineering manager at the Post Office Research Laboratory (then at Dollis Hill in London), saw the possibilities and persuaded others at the Post Office. His boss, Jack Tillman, tapped a new research fund of 12 million pounds to study ways to decrease fiber loss.

With Kao almost evangelically promoting the prospects of fiber communications, and the Post Office interested in applications, laboratories around the world began trying to reduce fiber loss. It took four years to reach Kao's goal of 20 dB/km, and the route to success proved different than many had expected. Most groups tried to purify the compound glasses used for standard optics, which are easy to melt and draw into fibers. At the Corning Glass Works (now

Corning, Inc.), Robert Maurer, Donald Keck, and Peter Schultz (Figure 1-11) started with fused silica, a material that can be made extremely pure, but has a high melting point and a low refractive index. They made cylindrical preforms by depositing purified materials from the vapor phase, adding carefully controlled levels of dopants to make the refractive index of the core slightly higher than that of the cladding, without raising attenuation dramatically. In September 1970, they announced they had made singlemode fibers with attenuation at the 633-nanometer (nm) helium neon line below 20 dB/km. The fibers were fragile, but tests at the new British Post Office Research Laboratories facility in Martlesham Heath confirmed the low loss.

The Corning breakthrough was among the most dramatic of many developments that opened the door to fiber optic communications. In the same year, Bell Labs and a team at the Loffe Physical Institute in Leningrad (now St. Petersburg) made the first semiconductor diode lasers able to emit carrier waves (CW) at room temperature. Over the next several years, fiber losses dropped dramatically, aided both by improved fabrication methods and by the shift to longer wavelengths where fibers have inherently lower attenuation.

Figure 1-11   Donald Keck, Robert Maurer, and Peter Schultz (left to right), who made the first low-loss fibers in 1970 at Corning. Courtesy Corning, Incorporated

Early singlemode fibers had cores several micrometers in diameter and in the early 1970s that bothered developers. They doubted it would be possible to achieve the micrometer-scale tolerances needed to couple light efficiently into the tiny cores from light sources or in splices or connectors. Not satisfied with the low bandwidth of step-index multimode fiber, they concentrated on multimode fibers with a refractive-index gradient between core and cladding, and core diameters of 50 or 62.5 micrometers. The first generation of telephone field trials in 1977 used such fibers to transmit light at 850 nm from gallium-aluminum-arsenide laser diodes.

Those first-generation systems could transmit light several kilometers without repeaters, but were limited by loss of about 2 dB/km in the fiber. A second generation soon appeared, using new indium gallium arsenide phosphide (InGaAsP) lasers that emitted at 1.3 micrometers, where fiber attenuation was as low as 0.5 dB/km, and pulse dispersion was somewhat lower than at 850 nm. Development of hardware for the first transatlantic fiber cable showed that singlemode systems were feasible, so when deregulation opened the long-distance phone market in the early 1980s, the carriers built national backbone systems of singlemode fiber with 1300-nm sources. That technology has spread into other telecom applications and remains the standard for most fiber systems.

However, a new generation of singlemode systems is now beginning to find applications in submarine cables and systems serving large numbers of subscribers. They operate at 1.55 micrometers, where fiber loss is 0.2 to 0.3 dB/km, allowing even longer repeater spacings. More important, erbium-doped optical fibers can serve as optical amplifiers at that wavelength, avoiding the need for electro-optic regenerators. Submarine cables with optical amplifiers can operate at speeds to 5 gigabits per second and can be upgraded from lower speeds simply by changing terminal electronics. Optical amplifiers also are attractive for fiber systems delivering the same signals to many terminals, because the fiber amplifiers can compensate for losses in dividing the signals among many terminals.

The biggest challenge remaining for fiber optics is economic. Today telephone and cable television companies can cost justify installing fiber links to remote sites serving tens to a few hundreds of customers. However, terminal equipment remains too expensive to justify installing fibers all the way to homes, at least for present services. Instead, cable and phone companies run twisted wire pairs or coaxial cables from optical network units to individual homes. Time will see how long that lasts.

REVIEW QUESTIONS

1. Confining light in a material by surrounding it by another material with lower refractive index is the phenomenon of _____
   a. cladding.
   b. total internal reflection.
   c. total internal refraction.
   d. transmission.

2. Abraham van Heel, in order to increase the total internal reflection, covered bare fiber with transparent cladding of _____
   a. higher refractive index.
   b. lower refractive index.
   c. higher numerical aperture.
   d. lower numerical aperture.

3. The high loss of early optical fiber was mainly due to _____
   a. impurities.
   b. silica.
   c. wave guides.
   d. small cores.

4. _____, using fused silica, made the first low loss (<20 dB/Km) singlemode optical fiber.
   a. Standard Telecommunications Laboratory
   b. The Post Office Research Laboratory
   c. Corning Glass Works
   d. Dr. Charles K. Kao

5. Erbium-doped optical fiber can serve as _____
   a. cladding.
   b. a pulse suppresor.
   c. a regenerator.
   d. an amplifier.

BASICS OF FIBER OPTICS

ELIAS A. AWAD

INTRODUCTION

Optical fiber is the medium in which communication signals are transmitted from one location to another in the form of light guided through thin fibers of glass or plastic. These signals are digital pulses or continuously modulated analog streams of light representing information. These can be voice information, data information, computer information, video information, or any other type of information.

These same types of information can be sent on metallic wires such as twisted pair and coax and through the air on microwave frequencies. The reason to use optical fiber is because it offers advantages not available in any metallic conductor or microwaves.

The main advantage of optical fiber is that it can transport more information longer distances in less time than any other communications medium. In addition, it is unaffected by the interference of electromagnetic radiation, making it possible to transmit information and data with less noise and less error. There are also many other applications for optical fiber that are simply not possible with metallic conductors. These include sensors/scientific applications, medical/surgical applications, industrial applications, subject illumination, and image transport.

Most optical fibers are made of glass, although some are made of plastic. For mechanical protection, optical fiber is housed inside cables. There are many types and configurations of cables, each for a specific application: indoor, outdoor, in the ground, underwater, deep ocean, overhead, and others.

An optical fiber data link is made up of three elements (Figure 2-1):

1. A light source at one end (laser or light-emitting diode [LED]), including a connector or other alignment mechanism to connect to the fiber. The light source will receive its signal from the support electronics to convert the electrical information to optical information.
2. The fiber (and its cable, connectors, or splices) from point to point. The fiber transports this light to its destination.
3. The light detector on the other end with a connector interface to the fiber. The detector converts the incoming light back to an electrical signal, producing a copy of the original electrical input. The support electronics will process that signal to perform its intended communications function.

The source and detector with their necessary support electronics are called the transmitter and receiver, respectively.

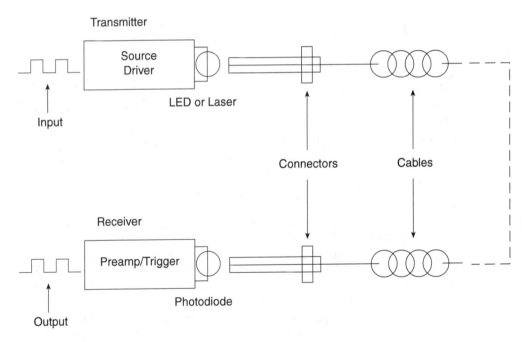

Figure 2-1 A typical fiber optic data link.

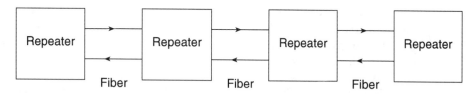

Figure 2-2   Long distance data links require repeaters to regenerate signals.

In long-distance systems (Figure 2-2) the use of intermediate amplifiers may be necessary to compensate for the signal loss over the long run of the fiber. Therefore, long-distance networks will be comprised of a number of identical links connected together. Each repeater consists of a receiver, transmitter, and support electronics.

OPTICAL FIBER

Optical fiber (Figure 2-3) is comprised of a light-carrying core surrounded by a cladding that traps the light in the core by the principle of total internal reflection. By making the core of the fiber of a material with a higher refractive index, we can cause the light in the core to be totally reflected at the boundary of the cladding for all light that strikes at greater than a critical angle. The critical angle is determined by the difference in the composition of the materials used in the core and cladding. Most optical fibers are made of glass, although some are made of plastic. The core and cladding are usually fused silica glass covered by a plastic coating, called the buffer, that protects the glass fiber from physical damage and moisture. Some all-plastic fibers are used for specific applications.

Glass optical fibers are the most common type used in communication applications. Glass optical fibers can be singlemode or multimode. Most of today's telecom and community antenna television (CATV) systems use singlemode fibers, whereas local area networks (LANs) use multimode graded-index fibers.

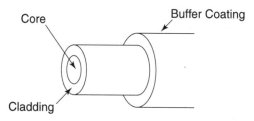

Figure 2-3   Optical fiber construction.

Singlemode fibers are smaller in core diameter than multimode fibers and offer much greater bandwidth, but the larger core size of multimode fiber makes coupling to low cost sources such as LEDs much easier. Multimode fibers may be of the step-index or graded-index design.

Plastic optical fibers are large core step-index multimode fibers, although graded-index plastic fiber is under development. Because plastic fibers have a large diameter and can be cut with simple tools, they are easy to work with and can use low-cost connectors. Plastic fiber is not used for long distance because it has high attenuation and lower bandwidth than glass fibers. However, plastic optical fiber may be useful in the short runs from the street to the home or office and within the home or office.

There are two basic types of optical fiber—multimode and singlemode (Figure 2-4). Multimode fiber means that light can travel many different paths (called modes) through the core of the fiber, entering and leaving the fiber at various angles. The highest angle that light is accepted into the core of the fiber defines

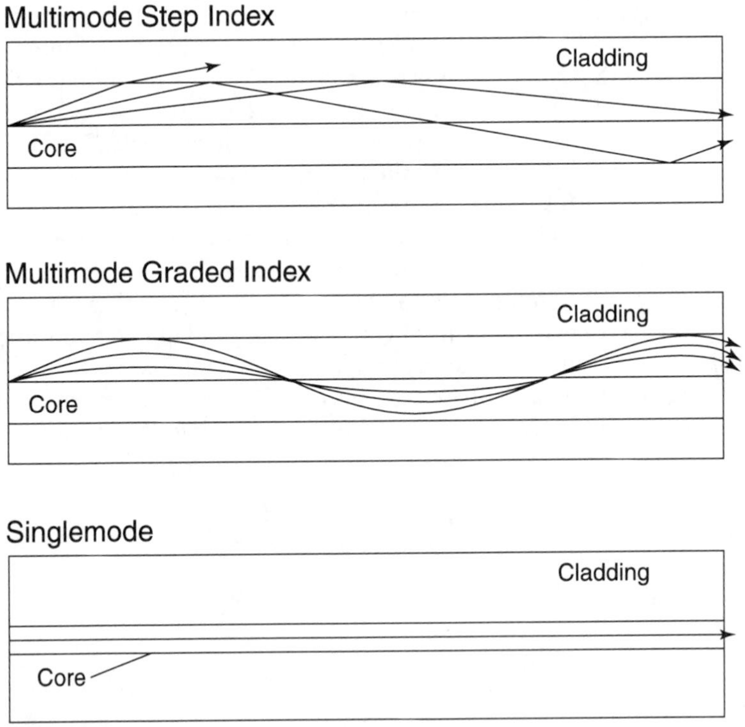

Multimode Step Index

Multimode Graded Index

Singlemode

Figure 2-4   The three types of optical fiber.

Table 2-1 Fiber Types and Typical Specifications

Fiber Type	Core/Cladding Diameter(m)	Attenuation 850 nm	Coefficient 1300 nm	(dBkm) 1550 nm	Bandwidth (MHz-km)
Multimode/Plastic	1 mm	(1 dB/m	@665 nm)		Low
Multimode/Step Index	200/240	6			50 @ 850 nm
Multimode/Graded Index	50/125	3	1		600 @1300 nm
	62.5/125	3	1		500 @1300 nm
	85/125	3	1		500 @1300 nm
	100/140	3	1		300 @1300 nm
Singlemode	8-9/125		0.5	0.3	high

the numerical aperture (NA). Two types of multimode fiber exist, distinguished by the index profile of their cores and how light travels in them (Table 2-1).

Step-index multimode fiber has a core composed completely of one type of glass. Light travels in straight lines in the fiber, reflecting off the core/cladding interface. The NA is determined by the difference in the indices of refraction of the core and cladding and can be calculated by Snell's law. Since each mode or angle of light travels a different path, a pulse of light is dispersed while traveling through the fiber, limiting the bandwidth of step-index fiber.

In graded-index multimode fiber, the core is composed of many different layers of glass, chosen with indices of refraction to produce an index profile approximating a parabola, where from the center of the core the index of refraction gets lower toward the cladding. Since light travels faster in the lower index of refraction glass, the light will travel faster as it approaches the outside of the core. Likewise, the light traveling closest to the core center will travel the slowest. A properly constructed index profile will compensate for the different path lengths of each mode, increasing the bandwidth capacity of the fiber by as much as 100 times over that of step-index fiber.

Singlemode fiber just shrinks the core size to a dimension about six times the wavelength of light traveling in the fiber and it has a smaller difference in the refractive index of the core and cladding, causing all the light to travel in only one mode. Thus modal dispersion disappears and the bandwidth of the fiber increases tremendously over graded-index fiber.

FIBER MANUFACTURE

Three methods are used today to fabricate moderate-to-low loss waveguide fibers: modified chemical vapor deposition (MCVD), outside vapor deposition (OVD), and vapor axial deposition (VAD).

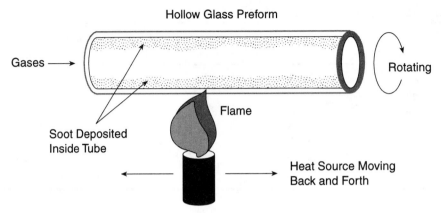

Figure 2-5 Modified chemical vapor deposition (MCVD).

Modified Chemical Vapor Deposition (MCVD)

In MCVD a hollow glass tube, approximately 3 feet long and 1 inch in diameter (1 m long by 2.5 cm diameter), is placed in a horizontal or vertical lathe and spun rapidly. A computer-controlled mixture of gases is passed through the inside of the tube. On the outside of the tube, a heat source (oxygen/hydrogen torch) passes up and down as illustrated in Figure 2-5.

Each pass of the heat source fuses a small amount of the precipitated gas mixture to the surface of the tube. Most of the gas is vaporized silicon dioxide (glass), but there are carefully controlled remounts of impurities (dopants) that cause changes in the index of refraction of the glass. As the torch moves and the preform spins, a layer of glass is formed inside the hollow preform. The dopant (mixture of gases) can be changed for each layer so that the index may be varied across the diameter.

After sufficient layers are built up, the tube is collapsed into a solid glass rod referred to as a preform. It is now a scale model of the desired fiber, but much shorter and thicker. The preform is then taken to the drawing tower, where it is pulled into a length of fiber up to 10 kilometers long.

Outside Vapor Deposition (OVD)

The OVD method utilizes a glass target rod that is placed in a chamber and spun rapidly on a lathe. A computer-controlled mixture of gases is then passed between the target rod and the heat source as illustrated in Figure 2-6. On each pass of the heat source, a small amount of the gas reacts and fuses to the outer surface of the rod. After enough layers are built up, the target rod is removed and the remaining soot preform is collapsed into a solid rod. The preform is then taken to the tower and pulled into fiber.

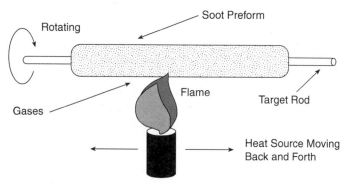

Figure 2-6 Outside vapor deposition (OVD).

Vapor Axial Deposition (VAD)

The VAD process utilizes a very short glass target rod suspended by one end. A computer-controlled mixture of gases is applied between the end of the rod and the heat source as shown in Figure 2-7. The heat source is slowly backed off as the preform lengthens due to tile soot buildup caused by gases reacting to the heat and fusing to the end of the rod. After sufficient length is formed, the target rod is removed from the end, leaving the soot preform. The preform is then taken to the drawing tower to be heated and pulled into the required fiber length.

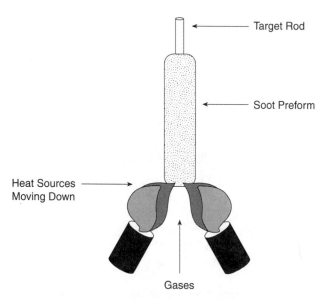

Figure 2-7 Vapor axial deposition (VAD).

Coating the Fiber for Protection

After the fiber is pulled from the preform, a protective coating is applied very quickly after the formation of the hair-thin fiber (Figure 2-8). The coating is necessary to provide mechanical protection and prevent the ingress of water into any fiber surface cracks. The coating typically is made up of two parts, a soft inner coating and a harder outer coating. The overall thickness of the coating varies between 62.5 and 187.5 μm, depending on fiber applications.

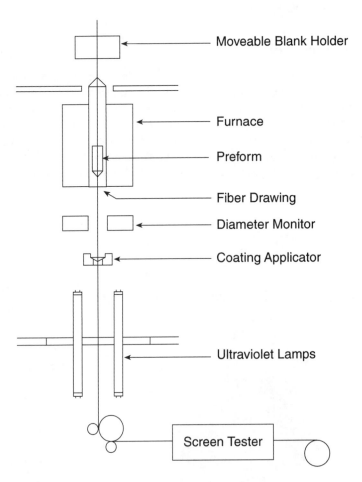

Figure 2-8 Drawing the fiber from the preform and coating the fiber.

These coatings are typically strippable by mechanical means and must be removed before fibers can be spliced or connectorized.

ADVANCED STUDY

What Is the Index of Refraction?

The index of refraction of a material is the ratio of the speed of light in vacuum to that in the material. In other words, the index of refraction is a measure of how much the speed of light slows down after it enters the material. Since light has its highest speed in vacuum, and since light slows down whenever it enters any medium (water, plastic, glass, crystal, oil, etc.), the index of refraction of all media is greater than one. For example, the index of refraction in a vacuum is 1, that of glass and plastic optical fibers is approximately 1.5, and water has an index of refraction of approximately 1.3

When light goes from one material to another of a different index of refraction, its path will bend, causing an illusion similar to the "bent" stick stuck into water. At its limits, this phenomenon is used to reflect the light at the core/cladding boundary of the fiber and trap it in the core (Figure 2-9). By choosing the material differences between the core and cladding, one can select the angle of light at which this light trapping, called total internal reflection, occurs. This angle defines a primary fiber specification, the numerical aperture.

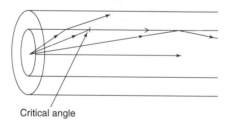

Critical angle

Figure 2-9 Total internal reflection in an optical fiber.

FIBER APPLICATIONS

Each type of fiber has its specific application. Step-index multimode fiber is used where large core size and efficient coupling of source power are more important than low loss and high bandwidth. It is commonly used in short, low-speed datalinks. It may also be used in applications where radiation is a concern, since it can be made with a pure silica core that is not readily affected by radiation.

Graded-index multimode fiber is used for data communications systems where the transmitter sources are LEDs. While four graded-index multimode fibers have been used over the history of fiber optic communications, one fiber now is by far the most widely used by virtually all multimode datacom networks—62.5/125 μm.

The telephone companies use singlemode fiber for its better performance at higher bit rates and its lower loss, allowing faster and longer unrepeated links for long-distance telecommunications. It is also used in CATV, since today's analog CATV networks use laser sources designed for singlemode fiber and future CATV networks will use compressed digital video signals. Almost all other high-speed networks are using singlemode fiber, either to support gigabit data rates or long-distance links.

FIBER PERFORMANCE

Purity of the medium is very important for best transmission of an optical signal inside the fiber. Perfect vacuum is the purest medium we can have in which to transmit light. Since all optical fibers are made of solid, not hollow, cores, we have to settle for second best in terms of purity. Technology makes it possible for us to make glass very pure, however.

Impurities are the unwanted things that can get into the fiber and become a part of its structure. Dirt and impurities are two different things. Dirt comes to the fiber from dirty hands and a dirty work environment. This can be cleaned off with alcohol wipes. Impurities, on the other hand, are built into the fiber at the time of manufacture; they cannot be cleaned off. These impurities will cause parts of optical signal to be lost due to scattering or absorption causing attenuation of the signal. If we have too many impurities in the fiber, too much of the optical signal will be lost and what is left over at the output of the fiber will not be enough for reliable communications.

Much of the early research and development of optical fiber centered on methods to make the fiber purity higher to reduce optical losses. Today's fibers are so pure that as a point of comparison, if water in the ocean was as pure, we would be able to see the bottom on a sunny day.

Optical glass fiber has another layer (or two) that surrounds the cladding, known as the buffer. The buffer is a plastic coating(s) that provides scratch protection for the glass below. It also adds to the mechanical strength of the fiber and protects it from moisture damage. On straight pulling (tension), glass optical fiber is five times stronger than some steel. But when it comes to twisting and bending, glass must not be stressed beyond its limits or it will fracture.

Fiber Attenuation

The attenuation of the optical fiber is a result of two factors—absorption and scattering (Figure 2-10). Absorption is caused by the absorption of the light and conversion to heat by molecules in the glass. Primary absorbers are residual OH^+ and dopants used to modify the refractive index of the glass. This absorption occurs at discrete wavelengths, determined by the elements absorbing the light. The OH^+ absorption is predominant, and occurs most strongly around 1000 nm, 1400 nm, and above 1600 nm.

The largest cause of attenuation is scattering. Scattering occurs when light collides with individual atoms in the glass and is anisotrophic. Light that is scattered at angles outside the critical angle of the fiber will be absorbed into the cladding or scattered in all directions, even transmitted back toward the source.

Scattering is also a function of wavelength, proportional to the inverse fourth power of the wavelength of the light. Thus, if you double the wavelength of the light, you reduce the scattering losses by 2^4 or 16 times. Therefore, for long-distance transmission, it is advantageous to use the longest practical wavelength for minimal attenuation and maximum distance between repeaters. Together, absorption and scattering produce the attenuation curve for a typical glass optical fiber shown in Figure 2-10.

Fiber optic systems transmit in the windows created between the absorption bands at 850 nm, 1300 nm, and 1550 nm, where physics also allows one to fabricate lasers and detectors easily. Plastic fiber has a more limited wavelength band that limits practical use to 660-nm LED sources.

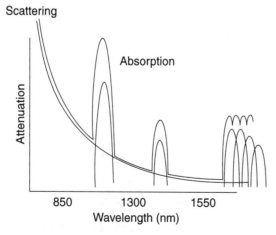

Figure 2-10   Fiber loss mechanisms.

Fiber Bandwidth

Fiber's information transmission capacity is limited by two separate components of dispersion: modal (Figure 2-11) and chromatic (Figure 2-12). Modal dispersion occurs in step-index multimode fiber where the paths of different modes are of varying lengths. Modal dispersion also comes from the fact that the index profile of graded-index multimode fiber is not perfect. The graded-index profile was chosen to theoretically allow all modes to have the same group velocity or transit speed along the length of the fiber. By making the outer parts of the core a lower index of refraction than the inner parts of the core, the higher order modes speed up as they go away from the center of the core, compensating for their longer path lengths.

Multimode Step Index

Multimode Graded Index

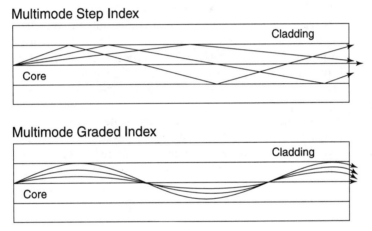

Figure 2-11 Modal dispersion, caused by different path lengths in the fiber, is corrected in graded-index fiber.

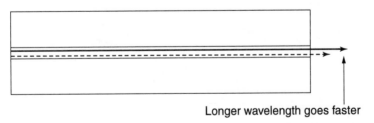

Longer wavelength goes faster

Figure 2-12 Chromatic dispersion occurs because light of different colors (wavelengths) travels at different speeds in the core of the fiber.

In an idealized graded-index fiber, all modes have the same group velocity and no modal dispersion occurs. But in real fibers, the index profile is a piecewise approximation and all modes are not perfectly transmitted, allowing some modal dispersion. Since the higher-order modes have greater deviations, the modal dispersion of a fiber (and therefore its laser bandwidth) tends to be very sensitive to modal conditions in the fiber. Thus the bandwidth of longer fibers degrades nonlinearly as the higher-order modes are attenuated more strongly.

The second factor in fiber bandwidth is chromatic dispersion. Remember, a prism spreads out the spectrum of incident light since the light travels at different speeds according to its color and is therefore refracted at different angles. The usual way of stating this is the index of refraction of the glass is wavelength dependent. Thus, a carefully manufactured graded-index multimode fiber can only be optimized for a single wavelength, usually near 1300 nm, and light of other colors will suffer from chromatic dispersion. Even light in the same mode will be dispersed if it is of different wavelengths.

Chromatic dispersion is a bigger problem with LEDs, which have broad spectral outputs, unlike lasers that concentrate most of their light in a narrow spectral range. Chromatic dispersion occurs with LEDs because much of the power is away from the zero dispersion wavelength of the fiber. High-speed systems such as Fiber Distributed Data Interface (FDDI), based on broad output surface emitter LEDs, suffer such intense chromatic dispersion that transmission over only 2 kilometer of 62.5/125 fiber can be risky.

Modal Effects on Attenuation and Bandwidth

The way light travels in modes in multimode fiber can affect attenuation and bandwidth of the fiber. In order to model a network or test multimode fiber optic cables accurately and reproducibly, it is necessary to understand modal distribution, mode control, and attenuation correction factors. Modal distribution in multimode fiber is important to measurement reproducibility and accuracy.

ADVANCED STUDY

What Is Modal Distribution?

In multimode fibers, some light rays travel straight down the axis of the fiber while all the others wiggle or bounce back and forth inside the core. In step-index fiber, the off-axis rays, called "higher-order modes," bounce back and forth from core/cladding boundaries as they are transmitted down the fiber. Since these higher-order modes travel a longer distance than the axial ray, they are responsible for the dispersion that limits the fiber's bandwidth.

In graded-index fiber, the reduction of the index of refraction of the core as one approaches the cladding causes the higher-order modes to follow a curved path that is longer than the axial ray (the "zero-order mode"). However, by virtue of the lower index of refraction away from the axis, light speeds up as it approaches the cladding, thus taking approximately the same time to travel through the fiber. Therefore the "dispersion," or variations in transit time for various modes, is minimized and bandwidth of the fiber is maximized.

However, the fact that the higher-order modes travel farther in the glass core means that they have a greater likelihood of being scattered or absorbed, the two primary causes of attenuation in optical fibers. Therefore, the higher-order modes will have greater attenuation than lower-order modes, and a long length of fiber that was fully filled (all modes had the same power level launched into them) will have a lower amount of power in the higher-order modes than will a short length of the same fiber.

This change in modal distribution between long and short fibers can be described as a "transient loss," and can make big differences in the measurements one makes with the fiber. It not only changes the modal distribution, it also changes the effective core diameter and apparent numerical aperture.

The term "equilibrium modal distribution" (EMD) is used to describe the modal distribution in a long fiber that has lost the higher-order modes. A "long" fiber is one in EMD, while a "short" fiber has all its initially launched higher-order modes.

In the laboratory, a critical optical system is used to fully fill the fiber modes and a "mode filter," usually a mandrel wrap that stresses the fiber and increases loss for the higher-order modes, is used to simulate EMD conditions. A "mode scrambler," made by fusion splicing a step-index fiber into the graded-index fiber near the source, can also be used to fill all modes equally.

When testing the network cable plant, using an LED or laser source similar to the one used in the system and short launch cables may provide as accurate a measurement as is possible under more controlled circumstances, since the LED approximates the system source. Alternately, one may use a mode conditioner (described below) to establish consistent modal distribution for testing cables.

Mode Conditioners

There are three basic "gadgets" used to condition the modal distribution in multimode fibers: mode strippers that remove unwanted cladding mode light, mode scramblers that mix modes to equalize power in all the modes, and mode filters that remove the higher-order modes to simulate EMD or steady-state conditions. These are discussed in Chapter 17.

REVIEW QUESTIONS

1. The main advantage(s) of optical is (are) its ability to _____ than other communications media.
   a. transport more information
   b. transport information faster
   c. transport information farther
   d. all of the above

2. A fiber optic data link is made up of three elements:
   1. _____
   2. _____
   3. _____

3. Plastic optical fibers are _____ fibers.
   a. singlemode
   b. large core step-index
   c. large core graded-index
   d. either a or b

4. Optical fiber is comprised of three layers:
   1. _____
   2. _____
   3. _____

5. What does 62.5 refer to when written 62.5/125?
   a. diameter of the core
   b. diameter of the cladding
   c. numerical aperture
   d. index profile

6. In graded-index optical fiber, the index profile approximates a parabola. The benefit of this is _____
   a. reduced bandwidth.
   b. reduced cross-talk.
   c. increased modal dispersion.
   d. reduced modal dispersion.

7. Three methods used to fabricate optical fiber:

    1. _____

    2. _____

    3. _____

8. Match the following fibers to the application they are best suited for:

    _____ Graded-index multimode   a. long-distance telecommunications

    _____ Step-index multimode   b. data communications

    _____ Singlemode   c. efficient source power coupling

9. The largest cause of attenuation is _____

    a. dopants.

    b. absorption.

    c. moisture.

    d. scattering.

10. Optical fiber's bandwidth, or information transmission capacity, is limited by two factors:

    1. _____

    2. _____

3

FIBER OPTIC NETWORKS

JIM HAYES AND PHIL SHECKLER

One often sees articles written about fiber optic communications networks that imply that fiber optics is "new." That is hardly the case. The first fiber optic telephone network was installed in Chicago in 1976, and by 1979, commercial fiber optic computer datalinks were available. Since then, fiber has become commonplace in the communications infrastructure.

If you make a long-distance call today, your voice is undoubtedly being transmitted on fiber optic cable, since it has replaced over 90 percent of all voice circuits for long-distance communications. Transoceanic links are being converted to fiber optics at a very high rate, since all new undersea cables are fiber optics. Phone company offices are being interconnected with fiber, and most large office buildings have fiber optic telephone connections into the buildings themselves. Only the last links to the home, office, and phone are not fiber.

CATV also uses fiber optics via a unique analog transmission scheme, but they are already planning on fiber moving to compressed digital video. Most large city CATV systems are being converted to fiber optics for reliability and in order to offer new services such as Internet connections and phone service. Only fiber offers the bandwidth necessary for carrying voice, data, and video simultaneously.

The LAN backbone also has become predominately fiber-based. The backend of mainframe computers is also primarily fiber. The desktop is the only holdout, currently a battlefield between the copper and fiber contingents.

Security, building management, audio, process control, and almost any other system that requires communications cabling have become available on fiber optics. Fiber optics really is the medium of choice for all high bandwidth and/or long-distance communications. Let us look at why it is, how to evaluate the economics of copper versus fiber, and how to design fiber networks with the best availability of options for upgradeability in the future.

IT IS REALLY ALL A MATTER OF ECONOMICS

The use of fiber optics is entirely an issue of economics. Widespread use occurred when the cost declined to a point that fiber optics became less expensive than transmission over copper wires, radio, or satellite links. However, for each application, the turnover point has been reached for somewhat different reasons.

Telephony

Fiber optics has become widely used in telephone systems because of its enormous bandwidth and distance advantages over copper wires. The application for fiber in telephony is simply connecting switches over fiber optic links (Figure 3-1). Commercial systems today carry more phone conversations over a single pair of fibers than could be carried over thousands of copper pairs. Material costs,

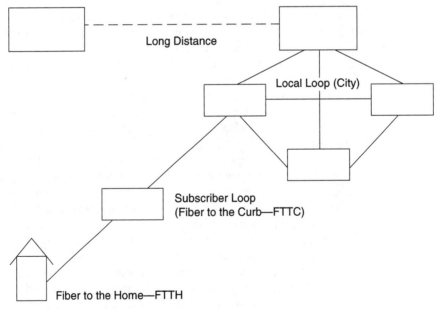

Figure 3-1 Telephone fiber optic architecture.

installation, and splicing labor and reliability are all in fiber's favor, not to mention space considerations. In major cities today, insufficient space exists in current conduit to provide communications needs over copper wire.

While fiber carries over 90 percent of all long-distance communications and 50 percent of local communications, the penetration of fiber to the curb (FTTC) and fiber to the home (FTTH) has been hindered by a lack of cost-effectiveness. These two final frontiers for fiber in the phone systems hinge on fiber becoming less expensive and customer demand for high bandwidth services that would be impossible over current copper telephone wires. Digital subscriber loop (DSL) technology has enhanced the capacity of the current copper wire home connections so as to postpone implementation of FTTH for perhaps another decade.

Telecommunications led the change to fiber optic technology. The initial use of fiber optics was simply to build adapters that took input from traditional telephone equipment's electrical signals on copper cables, multiplexed many signals to take advantage of the higher bit-rate capability of fiber, and used high-power laser sources to allow maximum transmission distances.

After many years of all these adapters using transmission protocols proprietary to each vendor, Bellcore (now Telcordia) began working on a standard network called SONET, for Synchronous Optical NETwork. SONET would allow interoperability between various manufacturers' transmission equipment.

However, the telephone companies' (telco's) transition to SONET was slow, a result of reluctance to make obsolete recently installed fiber optic transmission equipment and the slow development of the details of the standards. Progress has been somewhat faster overseas, where the equivalent network standard Synchronous Digital Hierarchy (SDH) is being used for first-generation fiber optic systems. SONET is now threatened by Internet protocol (IP) networks, since data traffic has surpassed voice traffic in volume and is growing many times faster, mostly due to the popularity of the Internet and World Wide Web.

CATV

In CATV, fiber initially paid for itself in enhanced reliability. The enormous bandwidth requirements of broadcast TV require frequent repeaters. The large number of repeaters used in a broadcast cable network are a big source of failure. And CATV systems' tree-and-branch architecture means upstream failure causes failure for all downstream users. Reliability is a big issue since viewers are a vocal lot if programming is interrupted!

CATV experimented with fiber optics for years, but it was too expensive until the development of the AM analog systems. By simply converting the signal from electrical to optical, the advantages of fiber optics became cost-effective. Now CATV has adopted a network architecture (Figure 3-2) that overbuilds the normal coax network with fiber optic links.

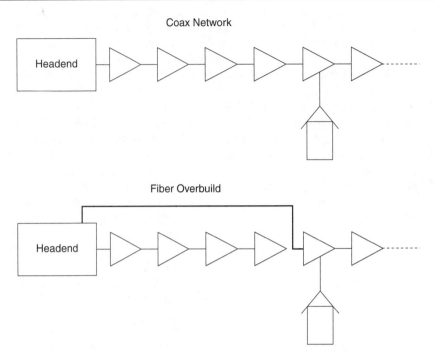

Figure 3-2 CATV architectures before and after fiber overbuild.

Fiber is easy to install in an overbuild, either by lashing lightweight fiber optic cable to the installed aerial coax or by pulling in underground ducts. The technology, all singlemode with laser sources, is easily updated to future digital systems when compressed digital video becomes available. The connection to the user remains coaxial cable, which has as much as 1 GHz bandwidth.

The installed cable plant also offers the opportunity to install data and voice services in areas where it is legal and economically feasible. Extra fibers can be easily configured for a return path. The breakthrough came with the development of the cable modem, which multiplexes Ethernet onto the frequency spectrum of a CATV system. CATV systems can literally put the subscriber on a Ethernet LAN and connect them to the Internet at much higher speeds than a dial-up phone connection. Adding voice service is relatively easy for the CATV operator as well.

Local Area Networks

For LANs and other datacom applications, the economics of fiber optics are less clear today. For low bit-rate applications over short distances, copper wire is undoubtedly more economical, but as distances go over the 100 meters called for in industry standards and speeds get above 100 Mb/s, fiber begins to look more attractive since copper requires more local network electronics and there are many problems installing and testing copper wire to high speed standards. Ability to upgrade usually tilts the decision to fiber since copper must be handled very carefully to operate at speeds where fiber is just cruising along.

Fiber penetration in LANs is very high in long-distance or high bit-rate backbones in large LANs, connecting local hubs or routers, but still very low in connections to the desktop. The rapidly declining costs of the installed fiber optic cable plant and adapter electronics combined with needs for higher bandwidth at the desktop are making fiber to the desk more viable, especially using centralized fiber architectures.

There are a large number of LAN standards today. The most widely used, called Ethernet or IEEE802.3 after its standards committee, is a 10, 100 MB/s or 1 GB/s LAN that operates with a protocol that lets any station broadcast if the network if free. Token ring (most often referred to as IBM Token Ring after its developer) is a 4 or 16 MB/s LAN that has a ring architecture, where each station has a chance to transmit in turn, when a digital "token" passes to that station. These two networks were developed originally based on copper wire standards. Fiber optic adapters or repeaters have been developed for these networks to allow using fiber optic cable for transmission where distance or electrical interference justifies the extra cost of the fiber optic interfaces for the equipment.

Most LANs have been designed from the beginning to offer the option of both copper wiring and fiber optics. Several of these networks were optimized for fiber. All share the common specification of speed: they are high-speed networks designed to move massive quantities of data rapidly between workstations or mainframe computers.

Fiber Distributed Data Interface (FDDI) is a high-speed LAN standard that was developed specifically for fiber optics by the ANSI X3T9.5 committee, and products are readily available. FDDI has a dual counter-rotating ring topology (Figure 3-3) with dual-attached stations on the backbone that are attached to both rings, and single-attached stations that are attached to only one of the rings through a concentrator. It has a token passing media access protocol and a 100-Mbit/s data rate. FDDIs dual ring architecture makes it very fault tolerant, as the loss of a cable or station will not prevent the rest of the network from operating properly.

ESCON (Figure 3-4) is an IBM-developed network that connects peripherals to the mainframe, replacing "bus and tag" systems. ESCON stands for Enterprise System Connection architecture. The network is a switched star architecture, using ESCON directors to switch various equipment to the mainframe computers. Data transfer rate started at 4.5 megabytes/second but was increased to 10 Mbytes/second. With an 8B/10B conding scheme, ESCON runs at about 200 Mbits/sec.

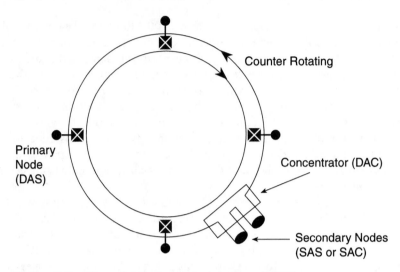

Figure 3-3   Fiber distributed data interface (FDDI).

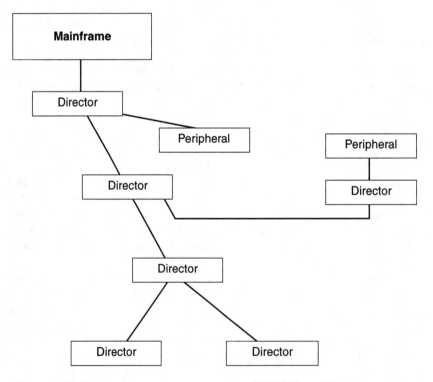

Figure 3-4   Enterprise system connection (ESCON) architecture.

Optically, ESCON and FDDI are similar. They use 1300-nm transmission for the higher bandwidth necessary with high-speed data transfer rates. Both single-mode and multimode cable plants are supported and distances up to 20 kilometers between directors.

Fibre Channel and High Performance Parallel Interface (HIPPI) are both high-speed links, not networks, that are designed to be used to interconnect high-speed data devices. The link protocol supports most fiber types and even copper cables for some short runs.

FIBER OR COPPER? TECHNOLOGY SAYS GO FIBER, BUT . . .

Fiber's performance advantages over copper result from the physics of transmitting with photons instead of electrons. Fiber optic transmission neither radiates radio frequency interference (RFI) nor is susceptible to interference, unlike copper wires that radiate signals capable of interfering with other electronic equipment. Because it is unaffected by electrical fields, utility companies even run power lines with fibers imbedded in the wires!

The bandwidth/distance issue is what usually convinces the user to switch to fiber. For today's applications, fiber is used at 100–200 Mb/s for datacom applications on multimode fiber, and telcos and CATV use singlemode fiber in the gigahertz range. Multimode fiber has a larger light-carrying core that is compatible with less expensive LED sources, but the light travels in many rays, called modes, that limit the bandwidth of the fiber. Singlemode fiber has a smaller core that requires laser sources, but light travels in only one mode, offering almost unlimited bandwidth.

In either fiber type, you can transmit at many different wavelengths of light simultaneously without interference; this process is called wavelength division multiplexing (WDM). WDM is much easier with singlemode fiber, since lasers have much better defined spectral outputs. Telephone networks using dense wavelength division multiplexing (DWDM) have systems now operating at greater than 80 MB/s. IBM developed a prototype system that uses this technique to provide a potential of 300 Gb/s on a LAN!

Which LANs Support Fiber?

That's easy, all of them. Some, such as FDDI or ESCON, were designed around fiber optics, whereas others, such as Ethernet or token ring, use fiber optic adapters to change from copper cable to fiber optics. In the computer room, you can get fiber optic channel extenders or ESCON equipment with fiber built in.

Where Is the Future of Fiber?

The future of fiber optics is the future of communications. What fiber optic offers is bandwidth and the ability to upgrade. Applications such as multimedia and video conferencing are driving networks to higher bandwidth at a furious pace. Over wide area networks, the installed fiber optic infrastructure can be expanded to accommodate almost unlimited traffic. Only the electronic switches need to be upgraded to provide orders of magnitude greater capacity. CATV operators are installing fiber as fast as possible since advanced digital TV will thrive in a fiber-based environment. Datacom applications can benefit from fiber optics also, as graphics and multimedia require more LAN bandwidth. Even wireless communications need fiber, connecting local low-power cellular or personal communication systems (PCS) transceivers to the switching matrix.

The Copper Versus Fiber Debate

Over the past few years, the datacom arena has been the site of a fierce battle between the *fiber* people and the *copper* people. First, almost 10 years ago, fiber offered the only solution to high-speed or long-distance datacom backbones. Although fiber was hard to install then and electrical/optical interfaces were expensive, when available at all, fiber was really the only reliable solution. This led to the development of the FDDI standard for a 100 Mb/s token ring LAN and the IBM ESCON system to replace bus and tag cables.

By 1989, FDDI was a reality, with demonstration networks operating at conferences to show that it really worked and that various vendors' hardware was interoperable. In 1990, IBM introduced ESCON as part of the System 390 introduction and fiber had become an integral part of their mainframe hardware. Everybody thought fiber had arrived.

However, at the same time, the copper wire manufacturers had developed new design cables that had much better attenuation characteristics at high frequencies. Armed with data that their Category 5 unshielded twisted pair (UTP) cables could transmit 100–150 Mb/s signals over 100 meters and surveys that showed that most desktop connections are less than that distance, they made a major frontal assault on the high-speed LAN marketplace. Simultaneously, other high-speed LAN standards, high-speed Ethernet and asynchronous transfer mode (ATM), which deliver FDDI speeds on copper wire, became popular. Now copper manufacturers are offering proprietary designs for copper cables that promise 250 MHz bandwidth, although the designs are years away from standardization. Many potential users continue to postpone making the decision to go to fiber.

So How Do You Decide Between Fiber and Copper?

Some applications are really black and white. Low bit-rate LAN connections at the desktop with little expectation of ever upgrading to higher bit rates should use copper. Long distances, heavy traffic loads, high bit rates, or high interference environments demand fiber. So if you have a backbone and Ethernet or token ring on the desktop, a fiber backbone and Category 5 UTP to the desktop makes good sense. If you already have a mainframe in the computer room and are using channel connections, you probably will use bus and tag cables for connections. But if you are extending those connections outside the computer room or buying a new mainframe, you will be getting fiber optic channel extenders or ESCON.

If either media will work in your application, it really comes down to economics—which solution is more cost-effective. But cost is a combination of factors, including system architecture, material cost, installation, testing, and "opportunity cost."

More end users are realizing that in a proper comparison, fiber right to the desktop can actually be significantly cheaper than a copper network. Look at the networks (Figure 3-5), and you will see what we mean.

The Traditional UTP LAN

The UTP copper LAN has a maximum cable length of 90 meters (about 290 ft.), so each desktop is connected by a unique UTP cable to a network hub located in a nearby "telecom closet." The backbone of the network can be UTP if the closets are close enough, or fiber optics if the distances are larger or the backbone runs a higher bandwidth network than can be supported on copper. Every hub connects to the main telecom closet with one cable per hub.

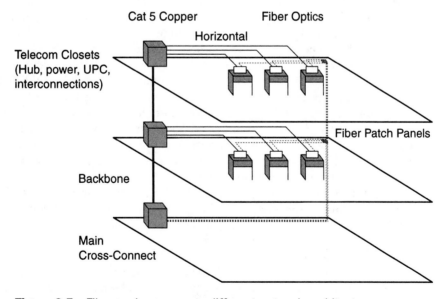

Figure 3-5   Fiber and copper use different network architectures.

In the telecom closet, every hub requires conditioned, uninterruptable power, since the network depends on every hub being able to survive a power outage. A data quality ground should be installed to prevent ground loops and noise problems. It will probably also have a rack to mount everything in (and the rack must be grounded properly.) Cables will be terminated in patch panels and patch cords will be used to connect cables to hubs.

The Fiber to the Desk LAN

Fiber optics is not limited in distance as is UTP cable. It can go as far as 2 kilometers (over 6,000 ft.), making it possible to bypass the local hubs and cable straight to the main telecom closet. It is likely there will be a small patch panel or wall box connecting desktop cables (probably zipcord) to a large fiber count backbone cable. At least 72 desktops can be connected on one backbone cable, which is hardly larger than one UTP cable.

So an "all-fiber" fiber network only has electronics in the main telecom closet and at the desktop—nothing in between. That means we do not need power or a UPS in the telecom closet—we do not even need a closet! Managing the network becomes much easier since all the electronics are in one location. Troubleshooting is simpler as well.

The Myth That Fiber Is More Expensive

The myth that fiber is more expensive has been copper's best defense against fiber optics. In a typical cost comparison, the architecture chosen is the typical copper one, and the cost of a link from the telecom closet to the desk, including electronics, is always higher for fiber—although by less and less each year.

But that is not a fair comparison! In a real comparison, we would price the complete networks shown in Figure 3-5. It would look more like Table 3-1.

So what happens if we total up the costs with this comparison? One estimate on a bank with no building construction costs had fiber costing only about $9 more per desktop. Another estimate had fiber costing only two-thirds as much as UTP. Several new construction projects claimed saving millions of dollars by eliminating all but one telecom closet in a large campus and thereby saving large amounts in building construction costs.

Fiber also saves money on testing. For fiber, it is a simple matter of testing the optical loss of the installed cable plant, including all interconnections to worldwide standards. The test equipment costs less than $1,000 and testing takes a few minutes per fiber.

Testing Category 5 or 6 UTP requires $3,000 to $50,000 in equipment and very careful control of testing conditions. Standards for testing are still continuously developed to keep up with new product development. If you consider the cost of testing, copper will probably cost a lot more than fiber!

Table 3-1. Comparison of Fiber and Copper Networks

	UTP Copper	Fiber
Desktop	Ethernet Network Interface Card for Cat 5	Ethernet Network Interface Card for fiber
Horizontal Cabling	Cat 5 cable, jacks, wall box, patch cord	Fiber zipcord, connectors, wall box, patch cord
Telecom Closet	Patch panel, patch cord, rack, hub, power connection, UPS, data ground	Wall mount patch panel
Backbone Cabling	One Cat 5 cable per connection	One multifiber cable per consolidation point
Main Telecom Closet	Patch panels, patch cords, electronics, power, UPC	Patch panels, patch cords, electronics, power, UPC
Building (relevant for new construction or major renovations)	Space for large bundles of cable, large floor or wall penetrations, big telecom closets, separate grounding for network equipment	Not needed

FUTURE-PROOFING THE INSTALLATION

As fast as networks are changing, always to higher speeds, *future-proofing* is a difficult proposition. When the decision to install fiber is made, follow up is needed in the planning phases to ensure that the best fiber optic network is installed. Planning for the future is especially important. You can easily install a cable plant for your LAN today that will fill your current needs and allow for network expansion for a long time in the future.

Follow industry standards such as EIA/TIA 568 and install a standard star architecture cable plant. Install lots of spare fibers since fiber optic cable is now inexpensive, but installation labor is expensive. Those extra fibers are inexpensive to add to a cable being installed today, but installing another cable in the future could be much more expensive.

What fibers should be installed? For multimode fibers, the most popular fiber today is 62.5/125 micron, since every manufacturer's products will operate optimally on this fiber. However, most equipment is also compatible with 50/125 fiber, which has already been installed in some networks, especially military and government installations in the United States and throughout Europe. All singlemode fiber is basically the same, so the choice is easier, although for most applications the specialty singlemode fibers (e.g., dispersion shifted or flattened) should be avoided.

Paying a premium for higher bandwidth or lower attenuation specifications in multimode fibers can allow more future flexibility. Very high-speed networks have forced fiber manufacturers to develop better fibers for gigabit networks. Installing that fiber today may make migrating to gigabit networks easier in the future.

How many fibers should be installed? Lots! Installation costs generally will be larger than cable costs. To prevent big costs installing additional cables in the future, it makes good sense to install large fiber count cables the first time; however, terminate only the fibers needed immediately, since termination is still the highest labor cost for fiber optics.

Backbone cables should include 48 or more fibers, half multimode and half singlemode. If you are installing fiber to the desktop, 12 fibers, again half and half, will provide for any network architecture now plus spares and singlemode fiber for future upgrades.

The new generation of gigabit networks may even be too fast for multimode fiber over longer distances and they will use lasers and singlemode fiber to achieve >1 GB/s data rates. If you want to use fiber for video or telecom, you may need the singlemode fiber now. But you may not want to terminate the singlemode fiber until you need it, since singlemode terminations are still more expensive than multimode; however, they are getting less expensive over time.

Fiber optics has grown so fast in popularity because of the unbelievably positive feedback from users. With proper planning and preparation, a fiber optic network can be installed that will provide the user with communication capability well into the next decade.

REVIEW QUESTIONS

1. Three areas in which fiber is used:
   1. _____
   2. _____
   3. _____

2. Match the application with the main reason fiber is the choice of transition medium.

   _____ LAN   a. upgradeability

   _____ CATV   b. reliability

   _____ Telecom   c. high bandwidth and distance advantages

3. FTTC stands for _____ .

4. FTTH stands for _____ .

5. The development of _____ made fiber cost-effective for CATV applications.
   a. repeaters
   b. FM systems
   c. AM analog systems
   d. enormous bandwidth

6. Match the following LAN standards with their counterparts in the right column.

   _____ Ethernet   a. dual counter-rotating ring
   _____ ESCON   b. most widely used LAN
   _____ FDDI   c. connects peripherals to a mainframe
   _____ Token ring   d. originally developed for copper networks

4

OPTICAL
FIBER CABLES

PAUL ROSENBERG

OPTICAL FIBER CABLE CONSTRUCTION

Because of the wide variety of conditions to which they are exposed, optical fibers have to be encased in several layers of protection. The first of these layers is a thin protective coating made of ultraviolet curable acrylate (a plastic), which is applied to the glass fiber as it is being manufactured. This thin coating provides moisture and mechanical protection.

The next layer of protection is a buffer that is typically extruded over this coating to further increase the strength of the single fibers. This buffer can be either a loose tube or a tight tube. Most data communication cables are made using either one of these two constructions. A third type, the ribbon cable, is frequently used in telecommunications (Figure 4-1).

Loose-tube (loose-buffer) cable is used mostly for long-distance applications and outside plant installations where low attenuation and high cable pulling strength are required. Several fibers can be incorporated into the same tube, providing a small-size, high-fiber density construction. The cost per fiber is also lower than for tight-buffered cables. The tubes may be filled with a gel or wrapped in an absorbent tape, which prevents water from entering the cable and offers additional protection to the fibers. Since these cables must be terminated either by fusion splicing to preconnectorized pigtails or by using breakout kits,

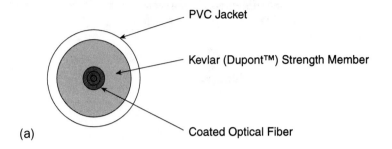

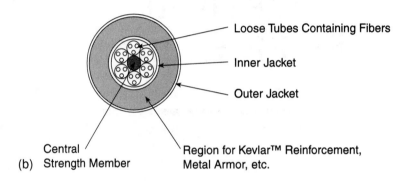

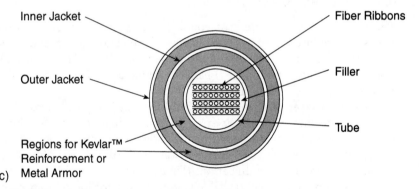

Figure 4-1 (a) Tight buffered fiber optic cable. (b) Loose-tube fiber optic cable. (c) Ribbon fiber optic cable.

they are more cost-effective for longer-distance applications than they are for short-distance applications. The fibers are completely separated from the outside environment. Therefore, the loose-tube cables can be installed with higher pulling tensions than tight-buffered cables.

A tight-buffered cable design is better when cable flexibility and ease of termination are a priority. Most inside cables are of the tight-buffered design because of the relatively short distances between devices and distribution racks. Military tactical ground support cables also use a tight-buffered design because of the high degree of flexibility required. A tight-buffered fiber can be cabled with other fibers, and then reinforced with Kevlar™, and jacketed to form a tightpack (distribution) cable. Another option is to individually reinforce each fiber with Kevlar, then jacket it. Several single fiber units can then be cabled together to obtain a breakout-style cable where each fiber can be broken out of the bundle and connectorized as an individual cable.

A ribbon-style cable consists of up to 12 coated fibers bonded to form a ribbon. Several ribbons can be packed into the same cable to form an ultra-high-density, low-cost, small-size design. Over 100 fibers can be put into a 1/2-inch square space with ribbon cables. Ribbon fibers can be either mass fusion spliced or mass terminated into array connectors, saving up to 80 percent of the time it takes to terminate conventional loose or tight-buffer cables.

Cable Jacketing

The materials used for the outer jacket of fiber optic cables not only affect the mechanical and attenuation properties of the fiber, but also determine the suitability of the cable for different environments, and its compliance to various National Electric Code (NEC) and Underwriters Laboratories (UL) requirements.

A cable that will be exposed to chemicals can utilize an inert fluorocarbon jacket such as Kynar, PFA, Teflon FEP, Tefzel, or Halar. These materials are suitable for a very wide range of applications, although they may be too stiff for some industrial applications.

Aerospace applications require that the cables be able to withstand a wide temperature range and be routed through the cramped environment of an aircraft. These cables are frequently rated for continuous operation from –65°C to +200°C, are less than 1/10 inch in size, and can sustain a bend radius of 1/2 inch.

Fire safety is a major issue. Cables used in an industrial environment, such as a power plant, are usually placed in horizontal trays. Several cable trays may be stacked in close proximity. In the event of a fire, both horizontal fire propagation and the ignition of lower cable trays by the dripping of flaming outer jacket material must be prevented. An irradiated Hypalon or XLPE jacket will meet the flame spread requirements (IEEE-383, 1974). When exposed to a flame, the jacket material will char rather than melt and drop burning material, thus preventing the ignition of cables in lower trays. Inside premises cables have to meet the requirements of the NEC Article 770. The outer jacket selection is essential to ensure compliance to the flame and smoke requirements.

Environmental and Mechanical Factors

Aside from buffer type, jacketing system, and flammability requirements, the cable design also must be based on the mechanical and environmental conditions that will be encountered throughout the system's life span.

A cable that will be pulled through conduits, ducts, or cable trays will have to incorporate a number of strength members and stiffening elements to add tensile strength and to prevent sharp bends from damaging the fibers. The addition of Kevlar increases the cable tensile strength. Kevlar can either be braided or longitudinally applied underneath the cable or fiber component jackets. The central strength member also serves both as a filler around which the fiber components

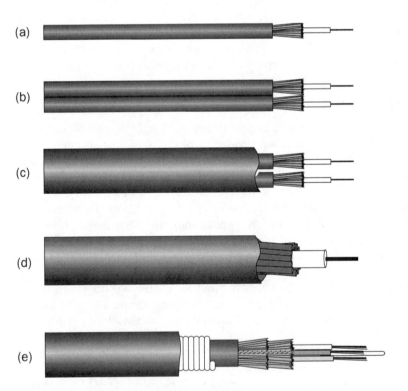

Figure 4-2   (a) Simplex cable. (b) Zipcord cable. (c) Tightpack cable. (d) Breakout cable. (e) Armored loose-tube cable.

are cabled and as a strength member when it incorporates steel, Kevlar, or epoxy glass rods. Another function of the epoxy glass central member is to act as an antibuckling component, counteracting the shrinkage of the jacketing elements at low temperatures and preventing microbends in the fibers. An epoxy glass rod central member should always be used in cables that may be exposed to temperatures below 0°C.

Industry Standards

Physical construction of optical cables is not governed by any agency. It is up to the designer of the system to make sure that the cable selected will meet the application requirements. However, five basic cable types (Figure 4-2) have emerged as de facto standards for a variety of applications.

1. *Simplex and zipcord:* One or two fibers, tight-buffered, Kevlar-reinforced and jacketed. Used mostly for patch cord and backplane applications (Figures 4-3 and 4-4).

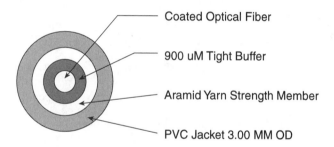

Coated Optical Fiber

900 uM Tight Buffer

Aramid Yarn Strength Member

PVC Jacket 3.00 MM OD

Figure 4-3   Simplex cable shown in cross-section.

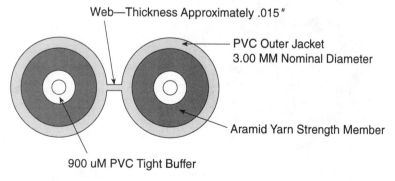

Web—Thickness Approximately .015″

PVC Outer Jacket
3.00 MM Nominal Diameter

Aramid Yarn Strength Member

900 uM PVC Tight Buffer

Figure 4-4   Zipcord cable shown in cross-section.

2. *Tightpack cables:* Also known as distribution style cables, consist of several tight-buffered fibers bundled under the same jacket with Kevlar reinforcement. Used for short, dry conduit runs and riser and plenum applications. These cables are small in size, but because their fibers are not individually reinforced, they need to be terminated inside a patch panel or junction box (Figure 4-5).

3. *Breakout cables:* Made of several simplex units cabled together. This is a strong, rugged design, and is larger and more expensive than the tightpack cables. Breakout cables are suitable for conduit runs and riser and plenum applications. Because each fiber is individually reinforced, this design allows for a strong termination to connectors and can be brought directly to a computer backplane (Figure 4-6).

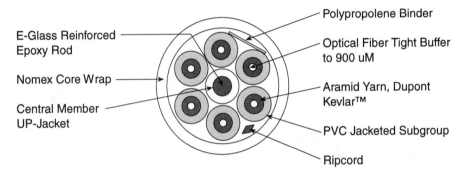

Figure 4-5 Tightpack cable shown in cross-section.

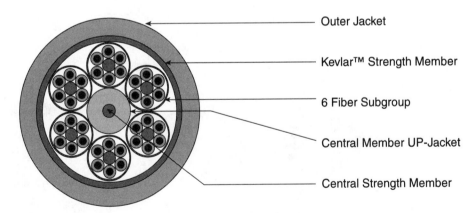

Figure 4-6 Breakout cable shown in cross-section.

4. *Loose-tube cables:* Composed of several fibers cabled together, providing a small, high-fiber count cable. This type of cable is ideal for outside plant trunking applications. Depending on the actual construction, loose-tube cables can be used in conduits, strung overhead, or buried directly in the ground (Figure 4-7).

5. *Hybrid or composite cables:* A lot of confusion exists over these terms, especially since the 1993 NEC switched its terminology from "hybrid" to "composite." Under the new terminology, a composite cable is one that contains a number of copper conductors properly jacketed and sheathed depending on the application, in the same cable assembly as the optical fibers. In issues of the code previous to 1993, this was called hybrid cable.

This situation is made all the more confusing because another type of cable is also called composite or hybrid. This type of cable contains only optical fibers but of two different types: multimode and single mode.

Remember that there is a great deal of confusion over these terms, with many people using them interchangeably. It is my contention that you should now use the term composite for fiber/copper cables, since that is how they are identified in the NEC. And, you should probably use hybrid for fiber/fiber cables, since the code does not give us much choice.

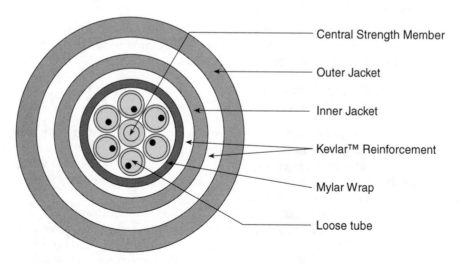

Figure 4-7 Loose-tube cable shown in cross-section.

CHOICE OF CABLES

The factors to be considered when choosing a fiber optic cable are:

1. Current and future bandwidth requirements
2. Acceptable attenuation rate
3. Length of cable
4. Cost of installation
5. Mechanical requirements (ruggedness, flexibility, flame retardance, low smoke, cut-through resistance)
6. UL/NEC requirements
7. Signal source (coupling efficiency, power output, receiver sensitivity)
8. Connectors and terminations
9. Cable dimension requirements
10. Physical environment (temperature, moisture, location)
11. Compatibility with existing systems

Composite Cables

If a system design calls for copper and fiber lying next to each other or in the same conduit, the designer should consider a composite cable. This would carry a number of copper conductors, properly jacketed and sheathed depending on the application, in the same cable assembly as the fiber optic cable.

Installation

Although the installation methods for both electronic wire cables and optical fiber cables are similar, there are two very important additional considerations that must be applied to optical fiber cables:

1. Never pull the fiber itself.
2. Never allow bends, kinks, or tight loops.

In order to keep these two rules, you must identify the strength member and fiber locations within the cables, then use the method of attachment that pulls most directly on the strength member. By paying careful attention to the strength limits and minimum bending radius limits and by avoiding scraping at sharp edges, damage can be avoided.

One guideline is that the pulling tension on indoor cables should never exceed 300 pounds. Another is that the minimum bending radius of an optical fiber cable should be no less than 10 times the cable diameter when not under tension, and 20 times cable diameter when being pulled into place (that is, 20 times cable diameter when under tension).

Cables in Trays

Optical fiber cables in trays should be carefully placed without tugging on the outer jacket of the cable. Care must be taken so that the cables are placed where they cannot be crushed. Flame retardant cables are recommended for interior installations.

Vertical Installations

Optical fibers in any type of vertical tray, raceway, or shaft should be clamped at frequent intervals, so that the entire weight of the cable is not supported at the top. The weight of the cable should be evenly supported over its entire length. Clamping intervals may vary from between 3 feet for outdoor installations with wind stress problems to 50 feet for indoor installations.

In such instances, the fibers sometimes have a tendency to migrate downward, especially in cold weather, which causes a signal loss (attenuation). This can be prevented by placing several loops about 1 foot in diameter at the top of the run, at the bottom of the run, and at least once every 500 feet in between.

Cables in Conduit

For all but the shortest pulls, loose-buffer cables are preferred, since they are stiffer and their jackets generally cause less friction than tight-buffered cables. Long pulls should be done with a mechanical puller that carefully controls pulling tension (Figure 4-8).

The cable lubricant must be matched to the jacket material of the cable. Most commercial lubricants will be compatible with popular types of cable jackets, but not in every case. Lubrication is considerably more important for optical fiber cables than for copper cables, since the fibers can be easily damaged.

Installation

In difficult installations, the cable-pulling force should be monitored with a tension meter. In these cases, the conduit should be prelubricated, and the cable lubricated also, as it is installed. Special lubricant spreaders and applicators are often used as well (Figure 4-9).

Except when tension meters are used, cable pulling should be done by hand, in continuous pulls as much as possible. Often this means pulling from a central manhole or pull box. During the pulling process, all tight bends, kinks, and twists must be carefully avoided. If they are not, the damaged cable may need to be removed and replaced with undamaged cable.

Two important devices to use when pulling optical fiber cables are swivel pulling eyes and breakaway swivels. The swivel pulling eyes allow the cable to turn independently of the pulling line or fish tape as it travels through the

Figure 4-8 For long pulls, the mechanical puller applies consistent tension and monitors it to prevent overstressing the fiber.

(a)   (b)

Figure 4-9   (a) Cable lubricant can be poured directly into the conduit before pulling. (b) For larger conduit, lubricant can be spread by pulling prepackaged bags through the conduit. Courtesy American Polywater Corporation

conduit. Since these cables are relatively fragile, the excessive twisting that could develop without the swivels should be carefully avoided. The breakaway swivel works in the same way as the swivel pulling eye, except that it will pull apart (thus stopping the pull) when the tension rises beyond a safe limit. In such a case, the cable must be pulled back out and reinstalled with more lubricant.

Attachment

The proper method of pulling optical fiber cables is to attach the pull wire or tape to the cable's strength member with the correct type of pulling eye (Figure 4-10).

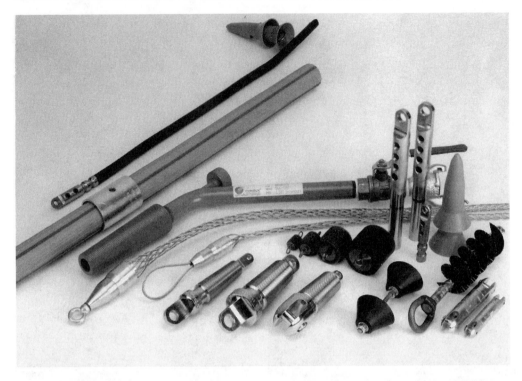

Figure 4-10   Numerous pulling eyes are available for various types of cable.

This avoids any tension on the fibers themselves. Unfortunately, it is not always easy to do.

When attaching to the strength members, the outer coverings are stripped back. Care must be taken not to damage the strength members, but stripping can normally be done with common tools. Kevlar or steel strength members can be tied directly to the pulling eye. Other more rigid types of strength members (such as fiberglass-epoxy) must be connected to a special set-screw device.

Indirect attachment can usually be well done with Kellems grips that firmly grip the cable jacket. For some larger cables, this type of attachment may actually be preferred. If you prestretch the Kellems grip and tape it firmly to the cable, much of the cable strain will be avoided.

Indirect attachment is not desirable when the fibers will be in the path of the forces between the pulling grip and the strength members. This is the case when the strength member is in the center of the cable, surrounded by the fibers. In such cases, only a small pulling force can be used.

Direct Burial

Generally, only heavy-duty cables can be directly buried. Numerous hazards affect directly buried optical fiber cables, such as freezing water, rocky soils, construction activities, and rodents (usually gophers). Burying the cables at least 3 or 4 feet deep avoids most of these hazards, but only strong metal braids or cables too large to bite will deter the gophers.

When plowing is used as an installation means, only loose-buffered cables are used, since they can withstand uneven pulling pressures better than tight-buffered cables. Where freezing water presents a problem, metal sheaths, double jackets, and gel fillings can be used as water barriers.

Installation

Rather than using expensive, heavy-duty cables, 1-inch polyethylene gas pipe is sometimes used to form a simple conduit. These tubes are also used as inner ducts, placed inside of larger (usually 4 inch) conduits. The plastic pipes provide a smooth passageway; by using several units inside of the larger conduit (with spacers holding them in place), the cables stay well organized. The plastic pipe can be smoothly bent, providing for very convenient installations and can reduce friction for easier and longer cable pulls.

Aerial Installations

When optical fibers are to be installed aerially, they must be self-supporting or supported by a messenger wire (See Article 321 of the NEC). Round, loose-buffer cables are preferred and should be firmly and frequently clamped or lashed to the messenger wire.

Cables for long outdoor runs are usually temperature stabilized. For the stabilization, steel is used if there are no lightning or electrical hazards. In other cases, fiberglass-epoxy is used. This type of dielectric cable is preferred for high vertical installations such as TV or radio towers.

Utilities use a special type of aerial cable called optical ground wire (OGW), which is a power cable capable of conducting high voltages with several fibers in the center. This type of power cable has gained acceptance with many power utilities that want communications fibers and prefer to install the OGW to get fiber capacity almost free.

Blown-in Fiber

Another method of installing fiber is to install special plastic tubes and *blow* the fibers in through the tubes using air pressure. This method does not use cable at all, merely buffered fibers. This method is not widely used and few installations of this type currently exist. However, it is becoming more popular since fibers can be easily removed and replaced for upgrades.

THE NATIONAL ELECTRICAL CODE

The requirements for optical fiber cable installation are detailed in Article 770 of the NEC. There are also alternate and/or supplementary requirements in the Life Safety Code.

Cable Designations

Remember that the NEC designates cable types differently than the rest of the trade. The code specifies horizontal cables, riser-rated cables, and plenum-rated cables. It also specifies cables as conductive or nonconductive. Note that a conductive cable is a cable that has any metal in it at all. The metal in a conductive cable does not have to be used to carry current; it may simply be a strength member.

All cables used indoors must carry identification and ratings per the NEC. Cables without markings should never be installed as they will not pass code!

NEC ratings are:

(OFN) Optical fiber nonconductive
(OFC) Optical fiber conductive
(OFNR) or (OFCR) Riser-rated cable for vertical runs
(OFNP) or (OFCP) Plenum-rated cables for installation in air-handling plenums

A legitimate question is whether an electrical inspector has any jurisdiction over installations that do not use conductive cables, the fact being that such cables do not carry any electricity. Nevertheless, such cables are dependent upon electronic devices to send and receive their signals. In addition, the NEC does address itself to all optical fiber cables.

Requirements

The main requirements of Article 770 are:

- When optical cables that have noncurrent-carrying conductive members contact power conductors, the conductive member must be grounded as close as possible to the point at which the cable enters the building. If desired, the conductive member may be broken (with an insulating joint) near its entrance to the building instead.
- Nonconductive optical cables can share the same raceway or cable tray with other conductors operating at up to 600 volts.
- Composite optical cables can share the same raceway or cable tray as other conductors operating at up to 600 volts.
- Nonconductive optical cables cannot occupy the same enclosure as power conductors, except in the following circumstances:

1. When the fibers are associated with the other conductors.
2. When the fibers are installed in a factory-assembled or field-assembled control center.
3. Nonconductive optical cables or hybrid cables can be installed with circuits exceeding 600 volts in industrial establishments where they will be supervised only by qualified persons.

- Both conductive and nonconductive optical cables can be installed in the same raceway, cable tray, or enclosure with any of the following:
  1. Class 2 or 3 circuits.
  2. Power-limited fire protective signaling circuits.
  3. Communication circuits.
  4. Community antenna television (CATV) circuits.
- Composite cables must be used exactly as listed on their cable jackets.
- All optical cables must be installed according to their listings. Refer to Section 770-53 to see the cable substitution hierarchy.

REVIEW QUESTIONS

1. Buffered fiber comes in three styles:
   1. _____
   2. _____
   3. _____

2. Loose-tube cable is used where _____
   a. ease of termination is a concern.
   b. high pulling strength is required.
   c. high flexibility is a concern.
   d. several fibers must fit in a small space.

3. A composite cable contains _____
   a. tight-buffered cables.
   b. singlemode and multimode fibers.
   c. loose-tube and tight-buffered fibers.
   d. copper conductors and optical fibers.

4. Match the type of cable listed with description in the right column.
   _____ Zipcord cable   a. contains single and multimode fibers
   _____ Tightpack cable   b. two fibers, tight-buffered, mostly used
   _____ Breakout cable   for patch cords
   _____ Loose-tube cable   c. contains copper conductors and optical
   _____ Composite cable   fiber
   _____ Hybrid cable   d. distribution cables
                                 e. a small diameter, high-fiber count cable
                                 f. several simplex units cabled together

5. When pulling fiber it is best to pull on the _____ of the cable.
   a. fiber
   b. buffer tubes
   c. jacket
   d. strength member

6. The minimum bending radius of an optical fiber cable should be no less than _____ times the cable diameter when being pulled into place.
   a. 10
   b. 15
   c. 20
   d. 25

5

SPECIFYING
FIBER OPTIC CABLE

ERIC PEARSON

CABLE PARAMETERS AND TYPICAL VALUES

In order to completely specify a fiber optic cable, you need to define at least 38 specifications. We divide these cable specifications into two subgroups, installation specifications and environmental, or long-term, specifications. Most of these specifications have a standard test technique by which the parameter is tested.

Note that not all specifications apply to all situations. You will need to review your application to determine which of the specifications in this section are needed. For example, cable installed in conduit or in protected locations will not need to meet crush load specifications.

INSTALLATION SPECIFICATIONS

The installation specifications are those that must be met in order to ensure successful installation of the cable. There are six such specifications:

1. Maximum recommended installation load, installation load, or installation force (in kg-force or pounds-force, or N)
2. Minimum recommended installation bend radius, installation bend radius, short-term bend radius, or loaded bend radius (in in. or mm)
3. Diameter of the cable

4. Diameter of subcable and buffer tubes
5. Recommended temperature range for installation (in degrees centigrade)
6. Recommended temperature range for storage (in degrees centigrade)

Maximum Recommended Installation Load

The maximum recommended installation load is the maximum tensile load that can be applied to a cable without causing a permanent change in attenuation or breakage of fibers. This characteristic must always be specified. It is particularly important in installations that are long, outdoors, or in conduits; it is of lesser importance when cables are laid in cable trays or installed above suspended ceilings. We present typical and generally accepted values of installation loads in Table 5-1. Choose the value that best fits your application.

If you believe that your application will require a strength higher than those typically specified, then you will want to specify a strength higher than those in Table 5-1. The cost increase of specifying such a higher strength is a small percentage, typically 5 to 10 percent, of the cost of the cable.

Minimum Recommended Installation Bend Radius

The minimum recommended installation bend radius is the minimum radius to which cable can be bent while loaded to the maximum recommended installation load. This radius is limited more by the cabling materials than by the bend radius of the fiber. This bending can be done without causing a permanent change in attenuation, breakage of fibers, or breakage of any portion of the cable structure. This bend radius is usually, but not always, specified as being no less than 20 times the diameter of the cable being bent. Specifying the bend radius is important when pulling by machine or hand through conduit, or in any long pulls.

Table 5-1   Typical Maximum Recommended Installation Loads

Application	Pounds Force
1 fiber in raceway or tray	67
1 fiber in duct or conduit	125
2 fiber in duct or conduit	
Multifiber (6–12) cables	250–500
Direct burial cables	600–800
Lashed aerial cables	>300
Self-support aerial cables	>600

In order to determine this value, you need to examine the locations in which you are to install your cable in order to determine the bend radius to which you will bend the cable during installation. Conversely, you can choose the cable and specify the conduits or ducts in which you are to install the cable so that you do not violate this radius.

Diameter of the Cable, Subcable, and Buffer Tubes

The cable must fit in the location in which it is to be installed. This is especially true if the cable is to be installed in a partially filled conduit. It will not be important if the cable is directly buried, installed above suspended ceilings, or in cable trays. If the diameter is limited by the space available, the diameter limits may be the only factor that determines which of the five designs of the cable you must choose. If cable diameter must be limited, the ribbon designs will be the smallest.

The diameter of the subcable and the buffer tube of the cable can also become a limiting factor. In the case of a "breakout" style cable, the diameter of the subcable must be smaller than the maximum diameter of the connector boot so that the boot will fit on the subcable. In addition, the diameter of the element must be less than the maximum diameter that the back shell of the connector will accept.

Recommended Temperature Ranges for Installation and Storage

All cables have a temperature range within which they can be installed without damage to either the cable materials or the fibers. It is more important for outdoor installations or in extreme (arctic or desert) environments and not important for indoor installations. In general, the materials of the cable restrict the temperature range of installation more than do the fibers. Note that not all cable manufacturers include the temperature range of installation in their data sheets. In this case, the more conservative temperature range of operation can be used.

In severe climates, such as those in deserts and the arctic, you will need to specify a recommended temperature range for storage (in degrees Centigrade). This range will strongly influence the materials used in the cable.

ENVIRONMENTAL SPECIFICATIONS

The environmental specifications are those that must be met in order to ensure successful operation of the cable in its environment. There are 21 such specifications.

1. Temperature range of operation
2. Minimum recommended long-term bend radius
3. Compliance with the NEC or local electrical codes

4. Long-term use load
5. Vertical rise distance
6. Flame resistance
7. UV stability or UV resistance
8. Resistance to damage from rodents
9. Resistance to damage from water
10. Crush loads
11. Resistance to conduction under high voltage fields
12. Toxicity
13. High flexibility/static versus dynamic applications
14. Abrasion resistance
15. Resistance to solvents, petrochemicals, and other chemicals
16. Hermetically sealed fiber
17. Radiation resistance
18. Impact resistance
19. Gas permeability
20. Stability of filling compounds
21. Vibration

Temperature Range of Operation

The temperature range of operation is the temperature range within which the attenuation remains less than the specified value. Typical ranges of operation are given in Table 5-2 for various types of applications. In general, there are very few applications in which fiber optic transmission cannot be used solely for reasons of temperature range of operation. In fact, some fibers have coatings that will survive continuous operation at 400°C. For operation at such high temperatures, fibers are usually, but not always, incorporated into a cable structure consisting of a metal tube. For operation at exceedingly low temperatures, cables are con-

Table 5-2 Typical Temperature Ranges of Operation

Application	Temperature Range (°C)
Indoor	−10 to +60, −10 to +50
Outdoor	−20 to +60,
	−40 to +50,
	−40 to +70
Military	−55 to +85
Aircraft	−62 to +125

structed of plastic materials that will retain their flexibility. For cables used at less severe temperatures (80–200°C), fluorocarbon plastics such as Teflon, Tefzel, Kynar, and others are used.

There are two reasons for considering the temperature range of operation: the physical survival of the cable and the increase of attenuation of the fiber when the cable is exposed to temperature extremes.

All cables are composed of plastic materials. These plastic materials have temperatures above and below which they will not retain their mechanical properties. After long exposure to high temperatures, plastics deteriorate, become soft, and, in some materials, crack. Under exposure to low temperatures, plastics become brittle and crack when flexed or moved. Obviously, under these conditions, the cable would cease to provide protection to the fiber(s).

The second reason for considering the temperature range of operation is the increase in attenuation that occurs when cables are exposed to extremes of temperature. Optical fibers have a sensitivity to being handled. This sensitivity is seen when the fibers are bent. This bending, which results in an increase in attenuation, is referred to as a "microbend-induced increase in attenuation." When a cable is subjected to temperature extremes, the plastic materials will contract and expand at rates much greater (100 times) than those rates of the glass fibers.

This contracting and expanding results in the fiber being bent on a microscopic level. Either the fiber is forced against the inside of the plastic tube as the plastic contracts, or the fiber is stretched against the inside of the tube as the plastic expands. In either case, the fiber is forced to conform to the microscopically uneven surface of the plastic. On a microscopic level, this is similar to placing the fiber against sandpaper. This microscopic bending results in light escaping from the core of the fiber. This escaping light results in an increase in attenuation. This type of behavior means that the user must determine the temperature range of operation in order to ensure that there will be enough light for the system to function properly.

Minimum Long-Term Bend Radius

The minimum recommended long-term bend radius is the minimum bend radius to which the cable can be bent for its entire lifetime. It is important for cables installed in conduits designed for electrical cables. It is usually, but not always, specified as being no less than 10 times the diameter of the cable.

Compliance with Electrical Codes

Fiber optic cables used in indoor applications must meet the requirements of the NEC and applicable local electric codes, some of which are more stringent than the NEC. Consult your local fire regulation authorities for those codes to which you must conform. Article 770 of the 1987 NEC addresses optical cables. Article 800 addresses cables that combine copper and fiber.

The NEC specifies six ratings. The first two letters in all ratings are "OF." The third is either an "N" or a "C." An "N" in the third place indicates a non-conductive, or all-dielectric design. A "C" in the third position indicates a cable containing conducting materials. The fourth letter, if any, indicates the rating. The least stringent are for "general use" cables, which must pass the UL 1581 test. Such cables are designated "OFN" or OFC."

Cables used in risers must not support the movement of fire from floor to floor. Such cables must pass the UL 1666 shaft test, which is more stringent than the UL 1581 test. Such cables are designated "OFNR" or "OFCR."

Cables installed in air-handling plenums must pass UL 910, the most stringent of the three tests. Such cables are designated "OFNP" or "OFCP" and must demonstrate adequate fire resistance and low smoke-generation characteristics. Use of plenum-rated cables allows you to reduce the total installed cost of the cables by eliminating the cost for the installation of metal conduit. The specification concerned with the requirements for plenum cables (both copper and fiber) is the NEC, Section 770. When choosing plenum-rated cables (OFNP or OFCP), consider plenum-rated PVC cables. These products have lower cost, easier installation, and better appearance than the original fluorocarbon cables.

Long-Term Use Load

Most fiber optic cables are designed for unloaded use, not for use with any substantial load. Substantial load occurs in applications such as vertical runs in elevator shafts, cables strung to elevators, cables placed on radio/TV towers, and cables strung outdoors between poles (aerial cables). In these cases, the cables are subjected to loads, either self-loads or loads from the environment, such as wind, snow, and ice loads on aerial cables. All of these factors depend on the spacing between poles.

Care in specifying the long-term use load characteristic is required to ensure that the strain the cable allows to be applied to the fiber(s) does not exceed a critical value. If this critical value is exceeded, it is likely that the fiber(s) will spontaneously, and for no apparent reason or cause, break. This value depends on the design and construction of the cable, but typically runs 10 to 30 percent of the maximum recommended installation load.

If the cable will experience a significant long-term use load, this specification will be more important than the maximum recommended installation load. Such cables, called "self-support" cables, are available from a number of manufacturers and are the cable of choice for use by power utilities for suspensions as long as 3,000 feet. In these cases, the maximum span length is specified instead of the long-term use load. Typical long-term use loads are presented in Table 5-3.

Table 5-3   Typical Maximum Recommended Use Loads

Application	Pounds Force
1 fiber in raceway or tray	23–35
1 fiber in duct or conduit	23–35
Multifiber (6–12) cables	33–330
Direct burial cables	132–180

Vertical Rise Distance

The vertical rise distance is related to the maximum use load. When cables are installed in a riser (within a building) or in a long vertical length (outdoors), the self-weight of the cable imposes a load on the cable. This load must be less than the maximum use load. Typical vertical rise distances are presented in Table 5-4.

Flame Resistance

Flame resistance is required for applications other than building applications, including shipboard and aircraft installations. In these applications, you will want to specify that the cables be constructed of flame-resistant materials. Many commonly used materials are either flame resistant in their most commonly used formulations, or can be made flame resistant through the use of additives. When you specify flame resistance, you will need to reference a specification, such as the UL specification 94, and specify the level of flame resistance required (i.e., V-0, V-1, V-2, etc.).

UV Stability or UV Resistance

If the cables are to be used continuously outdoors, then you need to specify that the cables be "UV resistant" or "UV stable." Otherwise, the cable jacket will crack and lose flexibility under exposure to sunlight. Most cables used continu-

Table 5-4   Typical Maximum Vertical Rise Distances

Application	Feet
1 fiber in raceway or tray	90
2 fiber in duct or conduit	50–90
Multifiber (6–12) cables	50–375
Heavy duty cables	1000–1640

ously outdoors have black polyethylene jacketing materials because this material has built-in UV-absorbing material and does not have plasticizers that evaporate over time. UV-resistant polyurethanes and polyvinyl chlorides (PVCs) are also available. However, the expected life of these two materials is much less than the more than 20-year life exhibited by polyethylene-jacketed telephone cables. Before choosing any jacket material other than black polyethylene for outdoor use, check its expected life span.

Resistance to Damage from Rodents

In environments containing active rodents, you will want to protect buried cable from damage caused by gnawing. There has been a trend away from the use of armored cables. Instead, buried inner ducts are used to provide the rodent resistance previously met by armored designs.

In some situations, you may need to specify the use of "armored" cables. This type of cable has an additional layer of material that acts to give the cable significant resistance to crushing and being bitten through. In addition, a final layer of plastic jacketing material is usually applied/extruded over the armor. There are penalties to these additional layers. First, armored cables are more expensive than nonarmored cables. Second, these cables are usually much less flexible than unarmored cables.

There are four basic types of armored cable products: galvanized steel armor (with or without plastic coating on the armor), copper tape armor, braided (stainless steel or bronze) armor, and dielectric armor. The armor most commonly used on fiber optic cables is galvanized steel. It is applied in a corrugated form or in a longitudinally welded/sealed form. It is effective and has the lowest cost of the armoring materials. However, it is the stiffest of the metallic armoring materials. Copper tape armor is helically wrapped around the cable with some spacing between the successive wraps. This type of product is rarely used on fiber optic cables. Because of its relatively flexible nature, braided armor is used in situations if rodent resistance and flexibility are required. Dielectric armoring is only available from a single source in the United States. This type of armoring is rarely needed and rarely used. It is the stiffest and most expensive of all types of armoring. The addition of a dielectric armor often doubles the cost of the cable.

Resistance to Damage from Water

If the cable is to be immersed in water, either permanently or for extended periods of time, as in most outdoor installations and all underwater installations, you will need to specify a "filled and blocked" cable. A filled and blocked cable has a filling material inside each of the loose buffer tubes and a blocking material that fills all empty space between the tubes. Failure to specify this type of cable will eventually result in an increase in attenuation and/or breakage of fibers. In addition, cables that are not filled and blocked can act as pipes by channeling water into electronic vaults

Some manufacturers supply "filled" cables. These cables are not as water-resistant as filled and blocked cables. Breakout cables are not filled and blocked. Before using any design that is not filled and blocked, request test data to support the water resistance claimed.

Crush Loads

The crush load is the maximum load that can be applied perpendicular to the axis of a cable without causing a permanent increase in attenuation or breakage of fibers. There are two crush loads: short-term and long term. Short-term can mean during installation or during use. The long-term crush load is that load that can be applied during the entire life of the cable.

Before you can determine the crushing requirements for your cable, you have to answer two basic questions. First, is the occurrence of crushing likely? If it is not a likely occurrence, then you will not need to be concerned with the crush performance of the cable you need. It has been the experience of the author that most of the cable products available today have crush performance sufficient to meet the needs of the typical user. This is so because most of the applications involve installation in relatively benign locations in which the occurrence of crushing is not likely. Examples of these benign locations include conduits, trays, cable troughs, plenums, and aerial locations. Examples of locations in which crushing performance is of importance are field tactical cables (in which the cable is likely to be run over by trucks and tanks), electronic news gathering (ENG), and temporary cable placement for sporting broadcast applications, shipboard use (in which the cable has a reasonable possibility of being crushed between bulkhead doors), and direct burial of fiber optic cable.

If you determine that crushing is of concern, then you need ask the second question: Is the application of a crush load likely to be a short-term or a long-term condition? If it is to be a short-term condition, then you will have two basic concerns: first, that the fiber not break; and second, that the "residual" or "hysteresis-type" increase in attenuation (which remains after the crush load is removed) be acceptable. Typical performances of commercial cables are given in Table 5-5.

Resistance to Conduction under High Voltage Fields

In a number of typical applications under high voltage fields, fiber optic cables need to be nonconducting. Some fiber optic cables in use are exposed to voltages as high as 1,000,000 volts. In other applications, fiber optic cables need to be unattractive to lightning. In these situations, you will specify that the cable be of an "all-dielectric construction." Such designs are commonly available.

Table 5-5 Typical Crush Strengths

Characteristic	Type of Cable	Pounds/Inch
Long-term crush load	>6 fibers/cable	57–400
	1–2 fiber cables	314–400
	Armored cables	450
Short-term crush load	>6 fibers/cable	343–900
	1–2 fiber cables	300–800
	Armored cables	600

Toxicity

Some applications—such as shipboard, aircraft, and mass transit—require "halogen-free" cables. These cables contain no halogens, which burn to produce acidic gases that attack lungs and corrode electronic equipment. These cables are 10 to 15 percent more expensive than PVC cables.

In addition to toxicity requirements, some municipalities require registration of all cables installed in order to keep track of the material content. In the United States, New York is the first state to require such registration. Cables manufactured for use in Japanese and European buildings are required to be halogen free.

High Flexibility/Static versus Dynamic Applications

In applications such as military field-tactical units and elevators, cables are subjected to repeated bending or flexing. In these applications, the cables need to meet a flexibility requirement. The need for high flexibility results in any of four requirements: flexure, high and/or low temperature bend, cable knot, and cable twist. Flexibility requirements must be met by both cable materials and by fibers.

Polyurethane jacketing materials are commonly used to meet this requirement. These materials will result in an increase in the cost of the cable, but will increase the flexibility to 10,000 cycles from the 1,000-cycle level available with the lower cost PVC and polyethylene jacketing materials.

Fibers can be made to meet the requirements of high flexibility and dynamic applications through the inclusion of a proof stress level. In such situations, as in elevator cables and in optical power ground wire (OPGW), some users have adopted a policy of requiring that the fibers be proof tested to at least 100 kpsi. Failures have been observed with dynamic loading of cables containing fibers proof stressed to only 50 kpsi.

Abrasion Resistance

In situations in which the cable is subject to abrasion, abrasion resistance must be specified. Need for this resistance will determine the material used as the jacket.

Resistance to Solvents, Petrochemicals, and Other Chemicals

In some situations, you need to specify that the cables be resistant to deterioration from exposure to certain chemicals. Examples to which cables are occasionally exposed are gasoline, aircraft fuel, fuel oil, greases, and crude oil. To ensure such resistance, an immersion test is required.

Hermetically Sealed Fiber

In applications requiring exposure of the cable to very high water pressures or high temperatures, the fiber must be hermetically sealed in order to retain its mechanical strength and/or its low attenuation. Hermetic sealing is required because contact with moisture (or other chemicals) results in significant reduction in the strength of the fiber, and absorption of hydrogen from water results in a significant increase in attenuation.

This hermetic sealing can be done in one of two methods. In the first method, the fiber is sealed inside of a welded steel tube. In the second method, the fiber is coated with a proprietary hermetic coating by the manufacturer. With both methods, the fiber is protected from degradation of its performance.

Radiation Resistance

When you intend to use a fiber optic cable in an environment subjected to ionizing radiation—such as in the core of a nuclear power plant, outer space, or an x-ray chamber—you must specify that both the cable materials and the fiber be radiation resistant. The cable materials must be radiation resistant in order to retain acceptable mechanical properties, since these properties tend to be degraded by exposure to ionizing radiation. The fiber must also be radiation resistant, since the attenuation of a fiber can be increased by such exposure.

Radiation-resistant fibers are available from a number of suppliers. Such fibers have smaller increases in attenuation (with increasing radiation dosage) than other more commonly used commercial fibers. In addition, these fibers have shorter recovery times and lower total residual increases in attenuation after such exposure.

Impact Resistance

In certain situations, you may want to specify the resistance of your fiber optic cable to impact forces. Examples of situations in which impact resistance is usually specified are cables used by military organizations in field tactical environments, cables being used in ENG applications, and any other situations in which heavy objects can be dropped on the cable. In these situations, you will specify impact resistance. When you do so, you will reference an Electronic Industries Alliance (EIA) RS 455 specification or a military specification.

As a practical matter, we have found most fiber optic cables to be highly resistant to damage from impacts. Unless impact is a likely occurrence in the environment in which the cable must survive, specification of impact resistance is not needed.

Gas Permeability

Some environments require that the cable not allow gases or moisture to travel through the cable. Examples of such environments are cables carrying signals from underground nuclear tests to equipment on the surface and underground cables leading to equipment located in underground vaults. In this case, gas or moisture permeability tests and limits must be specified.

Stability of Filling Compounds

Some environments subject the cable to frequent temperature and strain cycling. Such cycling has the potential to "pump" the filling compounds out of the ends of the cable. The pumping of filling compounds can cause problems to equipment at the ends of the cable. In this case, stability or flow tests and limits must be specified.

Vibration

In some situations, vibration may cause loose-tube cables to experience changes in attenuation. There is insufficient data available to recommend against loose-tube designs. However, in such situations, a tight-tube design may be preferable.

FOUR WAYS TO FUTURE-PROOF A SYSTEM

1. Include Spare Fibers in Cables

The U.S. ratio of currently used to total installed fibers is 1:4. Installing spare fibers offers two major advantages. First, you can use spare fibers in the event of a cable or connector problem. Second, spare fibers provide for future growth of fiber applications. Fiber is very inexpensive relative to installed cable cost and there is no cost for installing spare fibers as part of a cable being installed. If you need to install additional fibers in the future, you will incur two installation charges.

2. Include Singlemode Fibers in Multimode Cables

As bandwidths and bit rates increase, multimode fibers will eventually run out of capacity. Singlemode fibers provide essentially unlimited bandwidth.

3. Include Fibers in Any Copper Cables

Include fiber in any copper cable, as the cost of installing the fiber is free. You need not install connectors, although doing so is advised.

4. Use Dual Wavelength

Install 62.5/125 fibers that have been specified as dual wavelength, FDDI-grade fibers or better.

DESIGN SHORTCUTS

Fiber Choice

Multimode

Choose 62.5/125 μm fiber, the de facto standard for multimode fiber in the United States and the fiber specified by most network standards. Some other countries and some U.S. military applications use 50/125 μm, and new versions of 50/125 fiber are being developed for use with lasers in higher bandwidth systems such as 10 gigabit Ethernet.

Singlemode

Choose the 1300-nm singlemode fiber. Systems designed to operate at this wavelength have lower cost than 1550-nm systems. Do not choose fiber designed for both 1300 and 1550 nm unless you expect to use wavelength division multiplexing or optical amplifiers in the future.

Cable Design Choice

Indoor

1. For short distances [<1,200-1,335 feet], use breakout-type cable.
2. For longer distances, use premise-type cable.
3. If your environment is rugged, use breakout design; it is more rugged than the premise. The price premium is insurance against future maintenance cost.
4. Use all-dielectric design.
5. If plenum cables are required, look for plenum-rated PVC products.

Outdoor
1. Use one of three water-blocked and gel-filled loose-tube designs.
2. If fiber count is large [>36], compare the total installed cost of ribbon design to that of the other two loose-tube designs.
3. If midspan access is important, use the stranded loose-tube design.
4. Use all-dielectric design.

Indoor/Outdoor Cable Path

If cable path is both indoors and outdoors, you can eliminate a splice or connector pair by using an indoor/outdoor cable design. This design has an easily removable outdoor jacket over an inner structure that meets NEC requirements. Or, use a blocked cable that meets the appropriate NEC requirements.

Fiber Performance

Multimode

Choose dual wavelength specifications.

wavelength: 850/1300 nm
attenuation rate: < 3.75/1.0 dB/km
bandwidth-distance product: > 160/500 MHz-km
numerical aperture: .275 nominal (High bandwidth multimode fiber is becoming available to support new high-speed network such as Gigabit Ethernet.)

Singlemode

Choose single wavelength specifications.
wavelength: 1300 nm
attenuation rate: < 0.5 dB/km
dispersion: < 3.5 ps/km/nm @ 1310 nm

Cable Performance

Indoor

maximum recommended installation load: 360–500 pounds
temperature operating range: −10 to +60°C

Outdoor

maximum recommended installation load: 600 pounds
temperature operating range: −40 to +60°C
if rodent resistance required: armored or install in inner duct
strength members: epoxy fiberglass or flexible fiberglass
jacket: material black polyethylene

REVIEW QUESTIONS

1. Two specifications must be considered when specifying optical cable. They are:
   1. _____
   2. _____

2. A nonconductive optical fiber cable for use in an air-handling plenum would be labeled _____
   a. OFNP.
   b. OFCP.
   c. OFNR.
   d. OFCR.

3. Match items on the right with cables on left.
   _____ Minimum recommended installation bend radius
   _____ Minimum long-term bend radius
   _____ Plenum rated
   _____ "Self support" cable
   _____ Armored cable
   _____ "Filled and blocked" cable

   a. resistance to water damage
   b. OFCP
   c. rodent resistance
   d. no less than 20 time the diameter of the cable
   e. cable designed for long-term use loads
   f. no less than 10 times the diameter of the cable

4. Four ways to "Future-proof" an installation:
   1. _____
   2. _____
   3. _____
   4. _____

6

FIBER OPTIC CONNECTORS, SPLICES, AND TOOLS

JOHN HIGHHOUSE

FIBER JOINTS (CONNECTIONS)

For the purposes of this chapter we define a fiber joint as the point where two fibers are joined together to allow a light signal to propagate from one fiber into the next continuing fiber with as little loss as possible. Also, to keep from complicating procedures too greatly, all references are to glass fiber unless plastic fiber is specifically mentioned.

Although there are many reasons for fiber joints, the four most common are:

1. Fibers and cables are not endless and therefore must eventually be joined.
2. Fiber may also be joined to distribution cables and splitters.
3. At both transmit and receive termination points, fibers must be joined to that equipment.
4. The last and scariest reason is cable cuts and their subsequent restoration.

Since we have established a need for fiber joints, we should now make that task worthwhile. To that end, all fiber joints must be mechanically strong and optically sound with low loss. Fiber joints must be capable of withstanding moderate to severe pulling and bending tests. And, since the purpose of fiber is to transmit light, the fiber joint must transmit as much light power as possible with as little loss and back reflection as can be designed into the joint.

Fiber joints fall generally into two categories: the permanent or fixed joint that uses a fiber *splice*, and the terminating (nonfixed) joint that uses a fiber optic *connector*. Let us examine these individual types of joints.

Splices are used as permanent fixtures on outside and inside plant cables. Typical uses include reel ends, pigtail vault splices, and distribution breakouts. In addition to the benefits of low loss and high mechanical strength, additional considerations are expense per splice and possible reusability of the splice itself.

Fiber optic connectors are used as terminating fixtures for inside plant cables, outside plant cables as they terminate in a central office, interfaces between terminals on LANs, patch panels, and terminations into transmitters and receivers.

Whether one joins fibers using splices or connectors, one negative aspect is always common to both methods—signal loss. This loss of light power at fiber joints is called *attenuation*.

ATTENUATION

Attenuation is the loss of signal or light intensity as it travels through an optical fiber transmission system. Sometimes the losses occur in the fiber itself and other times at fiber joints. Measurement of attenuation loss is made in decibels (dB). The decibel is a mathematical logarithmic unit describing the ratio of output power to input power in any system (fiber or copper).

Attenuation in the optical fiber itself usually occurs as a result of absorption, reflection, diffusion, scattering, or dispersion of the photon packets within the fiber. However, losses also occur at splices and connections. The factors that cause attenuation in connectors or splices (Figure 6-1) fall into two categories: intrinsic and extrinsic losses.

Intrinsic losses occur from factors over which the craftsperson has very little control and are generally caused by engineering design or manufacturing flaws in the fiber itself. The more prominent intrinsic losses include:

1. Core eccentricity
2. Core ellipticity
3. Numerical aperture (NA) mismatch
4. Core diameter mismatch

Core eccentricity means that the exact center of the core center and the exact center of the cladding are not precisely the same, causing an overlap or underlap of fiber cores at a splice point. Core ellipticity (or ovality) is a departure from circularity. A very small variation in the roundness of a fiber core can affect the total system loss. Intrinsic loss through mismatch of NAs is not the fault of the craftsperson; however, care must be taken to butt the fibers as closely as possible

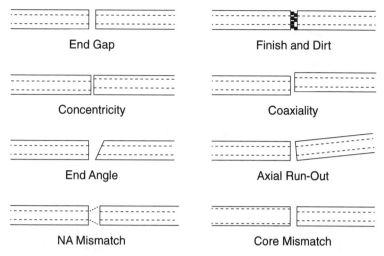

End Gap

Finish and Dirt

Concentricity

Coaxiality

End Angle

Axial Run-Out

NA Mismatch

Core Mismatch

Figure 6-1   Connector loss factors.

to counteract this mismatch. When splicing fibers having cores of different diameters, testing will show a significant loss when testing from the large core into the small core, and will show a supposed gain when testing from the small core into the large core.

Extrinsic losses, on the other hand, are caused by the mechanics of the joint itself. Frequent causes of extrinsic loss attenuation at splicing points include:

1. Misalignment of fiber ends caused by improper insertion techniques into splices and connectors.
2. Bad cleaves and poor polishing techniques resulting in poor end face quality.
3. Inadvertent air spaces between fibers at a splice or connection that have not been corrected with index-matching gel or liquid.
4. Contamination caused by dirt, wiping tissue, cotton swabs, shirt sleeves, or airborne dust particles. REMEMBER, IF YOU CAN SEE THE CONTAMINATION, IT'S TOO BIG FOR THE CORE TO PASS LIGHT THROUGH. See the section on cleaning connectors later in this chapter.
5. Another loss mechanism is *back reflection or reflectance* and is measured as optical return loss (Figure 6-2). As the light travels through the fiber, passing through splices and connections, finally arriving at the end point, some of that light is reflected back by fiber end faces at those manmade points. Optical return loss is generally only an issue with high-performance singlemode networks but is now also an issue with multimode networks used for gigabit networks.

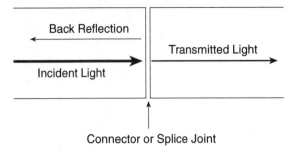

Figure 6-2 Optical return loss.

Typical allowable splice losses for singlemode fiber are 0 to 0.15 dB and with a return loss of better than 50 dB. In multimode fiber, typical splice losses are 0.0 to 0.25 dB, with an average of 0.20 dB and return loss of less than –50 dB. In the case of fiber connectors, singlemode allowable connector losses range from 0.1 to 1.0 dB per mated pair and return loss typically is less than –30 dB. Multimode connectors have a nominal connector loss of less than 0.75 dB per mated pair with a typical return loss better than 25 dB .

CONNECTORS

Remember that connectors are used as terminating fixtures for temporary non-fixed joints. As such, they are made to be plugged in and disconnected hundreds and possibly thousands of times. Since no one connector is ideal for every possible situation, a wide variety of connector styles and types have been developed over the short life of fiber communications. We can classify connectors by assigning them into five major categories:

1. Resilient ferrule
2. Rigid ferrule
3. Grooved plate hybrids
4. Expanded beam
5. Rotary

Of these types the rigid ferrule is by far the most common. Rigid ferrule types include the popular ST (compatible), FC, and SC, which use a single 2.5-millimeter cylindrical ferrule for fiber alignment. Other *simplex* connectors housing a single fiber, but no longer in common use today, include the SMA (905 and 906), D4, and the Biconic.

Duplex connectors contain two fibers allowing for a single connector body for both transmit and receive fibers. These connectors have come to the fore in recent years and are expected to gain popularity in the LAN arena. LAN hardware manufacturers have already adopted these connectors since they offer a much smaller size, allowing more links per panel space on network equipment.

Early examples of duplex connectors include the FDDI and ESCON. These connectors are rather large and cumbersome. Newer duplex connectors are designed to fit in the same work area outlet space as a standard RJ45 telephone jack and include the MT-RJ, Opti-Jack, and Volition connectors. These are commonly referred to as small form factor (SFF) connectors.

Although some SFF connectors are duplex designs, several others are miniature simplex connectors that are similar in design to the SC. The LC, LX-5, and MU connectors use smaller 1.25-millimeter ferrules and miniature bodies to allow twice the panel density of the earlier simplex connector designs. Examples of typical connector designs are shown in figure 6-3.

The end of an optical connector (Figure 6-4) can be either polished flat or with a PC finish, a slightly rounded, domed end to create a "physical contact," hence the PC designation. Physical contact of the fibers reduces the back reflection caused by air between the fiber ends. Some singlemode connectors may also have an "angled PC" (APC) finish. The ends are angled at 8 degrees to minimize back reflections at the point of connection. These connectors cannot be mated with the normal flat or domed polish types (Figure 6-4).

Although few, if any, of the original designs were compatible, nowadays compatibility exists between the same types from different manufacturers (i.e., ST or SC designs), thanks to marketplace pressures and standards committees. Although not compatible with all other connector styles, most ferrules are 2.5 millimeters and will loose fit for temporary testing purposes. For example, by lightly inserting the ferrule of an ST into an FC coupler, a "quick-and-dirty" test can be made for continuity. Hybrid adapters to allow coupling of different types of connectors are generally available as either sleeve connectors or patch cords. Although no single connector is best for every application, Table 6-1 lists the currently popular connectors found in many different types for various applications.

Choosing a Fiber Connector

With all of the myriad selections of connector types, styles, and physical characteristics available on the market, choosing the specific connector for your job is often a mystifying task. One important criterion is connector performance. When selecting a connector, comparisons of performance are generally based on:

- Insertion loss, usually 0.10 to 1.0 dB per connection
- Return loss (back reflection) varies from −20 (air gap like a SMA) to −60 dB (the best APC angle polished connectors)
- Repeatability of connection, usually specified at thousands of times

Your choice of fiber connector also may depend on whether you are mounting it onto singlemode or multimode fiber. Since singlemode connectors have a

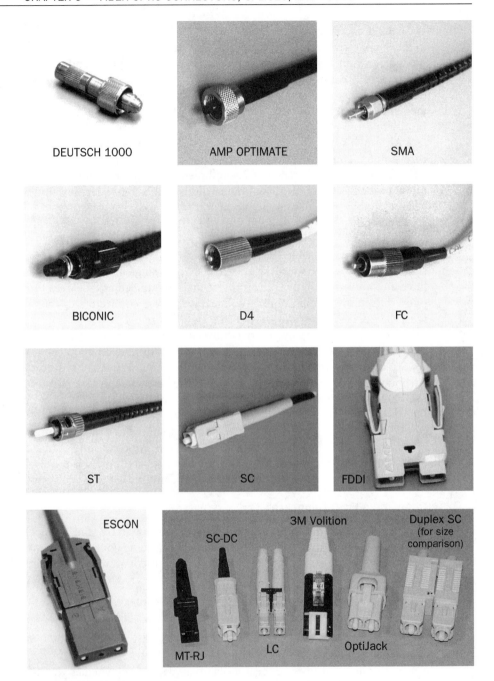

DEUTSCH 1000

AMP OPTIMATE

SMA

BICONIC

D4

FC

ST

SC

FDDI

ESCON

SC-DC

3M Volition

Duplex SC
(for size comparison)

MT-RJ

LC

OptiJack

Figure 6-3 Connector styles.

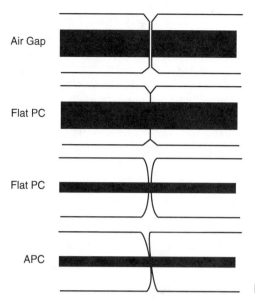

Air Gap

Flat PC

Flat PC

APC

Figure 6-4 Connector and finishes.

much tighter tolerance than multimode connectors, they may be used on either type of fiber. However the reverse is not true, that is, one may not use multimode connectors on single mode fiber because the loose tolerance will cause high loss with the very small singlemode core size. Generally multimode connectors are fitted onto multimode fibers because they are less precise and cost about one-half to one-third the cost of single mode connectors.

The accessibility of the fiber to casual users may cause you to anticipate rough handling. In this case, gripping strength of the connector on the cable becomes

Table 6-1 Popular Connectors for Data Communications and Telecommunications

Data communications (Mostly multimode)	Telecommunications (Mostly singlemode)
SMA (obsolete)	Biconic (obsolete)
ST (most widely used)	D4 (fading)
SC (for newer systems)	FC/PC (widely used)
FDDI (duplex)	SC (growing)
ESCON (duplex)	ST (singlemode version)
MT-RJ (new SFF duplex style)	LC (new SFF)
Volition (new SFF duplex style)	MU (SFF, outside United States)
Opti-Jack (new SFF duplex style)	

important to avoid pullouts by users. Gripping points of the connector may include the fiber itself, the primary plastic buffer coating (tight buffer), the loose-tube buffer, the cable strength members (Kevlar), and/or the cable jacket itself.

Another reason for choosing a particular type of connector is the type of equipment already purchased or currently in use. If, for instance, you are adding to an existing system already equipped with ST connectors, you should continue to use ST connectors to ensure compatibility systemwide. If you are using previously purchased electronics with Biconic connectors installed, then that will be your choice, unless, of course, you want to change all of the connections on the patch panels and electronics!

Finally, your choice may be influenced by industry standards or new developments in the marketplace. The Electronic Industries Alliance/Telecommunications Industry Association (EIA/TIA) standards for premises cabling calls for the SC connector, although they are considering the new SFF connectors. Many of the newer connectors offer the promise of lower cost or higher performance, which can also influence the decision.

Cable Termination and Connector Installation

Fiber optic connectors can be installed directly on most fiber optic cables, as long as the fiber has a tight buffer or jacket to protect it. Before the installation of connectors onto a loose-tube or ribbon fiber optic cable, a breakout kit may have to be installed. This procedure is not necessary on breakout cables having 3-millimeter jacketed fibers, but will be required on 250-, 500-, and some 900-micron tight-buffer cables. The breakout kit consists of plastic tubing into which the bare fibers are inserted to provide handling protection and strength when mounted onto connectors.

Installing a fiber connector onto a fiber is a widely varied process. The most common mounting methods are:

- Adhesives to hold the fiber in the connector and polished ferrules
   Epoxy glue with room temperature or oven cure
   Quick curing adhesives
   Hot Melt, preloaded adhesive (Hot Melt is a 3M trademark)
- Crimping to hold the fiber, with or without requiring polishing of the ferrule end
- Prepolished ferrules with fiber stub; connector is spliced onto the fiber.

The epoxy/polish method (Figure 6-5) is the oldest of all methods and is used today in all manufacturing plants and many field installations. This process involves filling the connector with a premixed two-part epoxy. The prepared and cleaned fiber is then inserted into the connector, which is crimped onto the cable. After curing the epoxy in an oven for the proper time (5 to 40 minutes) or overnight at room temperature, the end of the fiber is polished. The fiber must be

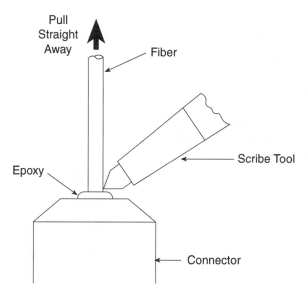

Figure 6-5 The fiber is epoxied into the connector body, then scribed and broken off just above the connector face before polishing.

scribed and cleaved nearly flush with the end of the connector and polished with a two or three fine lapping papers (Figure 6-6). The cleaved fiber is usually removed gently with hand-held film of about 12-micron finish in a process called "air polishing." Final polish papers start at 3 microns and go as fine as 0.3-micron grit. High volume terminations are usually lapped on polishing machines (Figure 6-7) that can handle anywhere from one to a dozen connectors simultaneously.

The Hot Melt (trademark of 3M) uses an adhesive preloaded into the connector. The connector is placed into an oven to soften the glue and allow insertion of the prepared fiber. After cooling, the scribe and polish process is the same as previously described.

Quick-cure adhesives include one- and two-part adhesives that cure in less than one minute. Many different adhesives are used for fiber termination, but it is important to not just use any quick-curing adhesive. The adhesive must meet stringent requirements for adhesion to the fiber and resistance to moisture or temperature extremes.

Some connectors, such as the 3M "CrimpLok" connector use no adhesive to capture the fiber. Instead, an internal malleable metal V-groove plate is locked on to the fiber holding it in place in the ferrule. The fiber cable is then affixed to the backbone of the connector by crimping. The connector requires a special fiber optic cleaver to prepare the fiber and a special polishing procedure.

Figure 6-6   A polishing puck holds the connector properly for polishing.

Impact mounting, as used by the Valdor connector, is another termination method that requires no adhesive. This connector uses a special metal ferrule rather than ceramic. After the fiber is stripped and cleaned, it is inserted into the

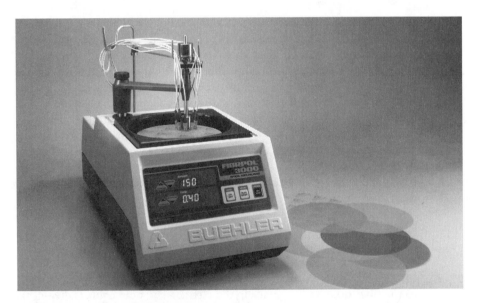

Figure 6-7   Automatic polishers can polish large quantities of connectors quickly. Courtesy Buehler LTD

connector and a hollow tool impacts the end face of the ferrule, swaging connector onto the fiber. The fiber stub is then scribed off and the end face is polished on a glass plate to provide a flat finish.

Another type of connector crimps the fiber to hold it in place, and then uses a special tool to cleave the fiber flush with the end face of the ferrule. These connectors require no polishing, so they are quickly terminated, but they typically have higher loss than polished connectors.

Cleave and crimp connectors, on the other hand, do not require any type of polish procedure and can be terminated very quickly (Figure 6-8). They already have a polished ferrule tip and are spliced to the fiber. They require only the insertion of a properly cleaved fiber to butt against the internal fiber stub. Once in place, the fiber connector is crimped to hold the fiber. These connectors often have higher loss than polished connectors, since they include both a connection and a splice and require more expensive tools for termination.

Each mounting method has its advantages and disadvantages, varying from ease of installation to cost per connector to performance qualities. See the section at the end of this chapter that compares actual termination procedures for several types of connectors.

Strip, Clean, and Cleave

The three basic steps for any fiber joint, whether splicing or connectorizing, are *strip, clean, and cleave*. Stripping involves the removal of the 250-micron primary coating and any other layers of protection on the individual fiber. Sometimes this protection takes the form of a 900-micron tight-buffer coating such as is found in indoor riser cables. The strip process must be accomplished using the correct stripping tool of the correct size. When stripping 250-micron coated loose-tube fiber, the entire length of fiber (usually no more than 3 inches) may be stripped in one pass with the tool. However, stripping 900-micron tight buffer will require that no more than 1/4 inch of tight buffer be removed at a time to prevent breaking the fiber. To avoid microbends and fiber stress, always use the tool at a near right angle and never wrap the fiber around your fingers to "get a better grip."

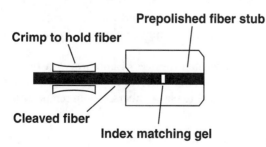

Prepolished fiber stub

Crimp to hold fiber

Cleaved fiber

Index matching gel

Figure 6-8   Cleave and crimp connectors have a short fiber already glued in the ferrule and polished.

The cleaning process is also an inspection and testing process. Once the fiber is stripped it must be cleaned using a lint-free wipe and reagent-grade isopropyl alcohol. The wipe is moistened and the fiber is pinched tightly and wiped using a curling motion. This will cause a fiber that has been scratched or cracked by the stripping process to break. Better now than later! Cleaning should be done in one pass if possible to minimize fiber handling. Make that wiping pad squeak to be sure that all stripped plastic residue is removed!

Cleaving (Figure 6-5) will take place using either a cleaver or a scribe and break process. In either case the fiber should end up with a cleave as near to 90 degrees as possible. When splicing using a cleaver, do not be tempted to clean the fiber again using a wipe as this will draw small glass particles and dirt to the end face. Instead, use a small piece of plastic tape to remove any remaining end-face contamination by performing the "tape-tap" procedure.

Fiber Optic Adhesives

Adhesives have been used since the onset of fiber optics to affix most connectors. The primary purpose is to hold the fiber in place and prevent any movement (pistoning). The adhesives can also supply support and strength to the fiber, specifically at the connector end. Adhesives are also being used to hold protective boots to the fiber jacket. Similar materials are also being used for laser applications. This discussion focuses on the traditional application of attaching fiber into a connector.

Enormous strides have been made in the past 15 years to keep pace with the high performance demands of the latest fiber connectors. New increasingly rigorous aging tests require formulated adhesives that allow little or no dimensional movement of the fiber in the ferrule over strenuous conditions. These systems must be rigid enough for polishing and yet flexible enough to withstand differences in expansion rates from the wide variety of substrates used in fiber optic connectors and cables. Of course, they also must maintain a good bond to all of these substrates.

Application techniques of epoxies vary depending on the type of connector and the production requirements. Most common is the injection of adhesive into the connector using a syringe or automated cartridge. Applying the adhesive directly to the fiber is also used. However, to provide the best bond strength, the adhesive must wet out the surfaces of the fiber and the connector ferrule sufficiently. It is more difficult to achieve the maximum physical properties using the latter method.

There is a misconception concerning the use of epoxies with plastic fibers. Epoxies work well with both plastic and glass fibers. Formulations currently available will not "attack" or contaminate plastic fibers. Room temperature-curing

---

This section was contributed by Barry Siroka, formerly Photonics Business Manager, Tracon.

epoxies are usually used for plastic fibers, although heat-curing and fast-gelling epoxies have been used successfully.

Three basic types of epoxies are currently available for fiber optic connectors. Heat-curing, room temperature-curing, and fast-gelling epoxies can all be used in most connectors with any type of fiber.

Heat-curing epoxies have the highest temperature capabilities of all epoxy systems. These systems are primarily used in connectors where fast cure is desired. It is usually recommended that these systems reach a temperature of 90–100°C in order for the chemical reaction between the epoxy and the hardener to take place. Some heat-curing epoxies can cure in as quickly as one minute at 150°C. An added benefit to this high temperature curing requirement is that these systems usually have the longest working life. They can also be color coded to ensure a proper mix, and some will change color upon cure.

The second type of epoxy used in fiber optic connectors are the room temperature-curing (RT) systems. These are the most popular and can be used in singlemode and multimode connectors. These systems will cure with no heat overnight to a tack-free surface. Many fabricators will speed up the cure by heating the connectors up to 65°C. Full cure is in 1 hour with sufficient cure occurring in 15 minutes for polishing. Some systems can be completely cured at 90°C in 10 minutes. RT-curing systems have a working life of 15 to 60 minutes.

Fast-gelling epoxies are used primarily in field installations where no power is available for ovens and/or speed is essential. These systems have a dual cure mechanism that allows for the fast gel (and therefore can be polished in as little as 10 minutes). After gelling, they will complete their cure overnight at room temperature. Properly applied formulations have been shown not to piston after curing.

Fast-gelling epoxies traditionally have an approximate working time of 5 minutes. New variations can offer a 10-minute working time. These new variations allow for less waste as there is more time to use all the material from one mix. These versions can be polished in as quickly as 20 minutes and will also complete their cure in 12 to 18 hours.

In dealing with epoxies, do not overlook safety issues. Many chemicals can cause dermatitis or respiratory ailments. Therefore, when handling any adhesive product, it is always recommended that care be taken to prevent contact with the skin and adequate ventilation should be employed. The hardeners are usually the most offensive. Again, prepackaged epoxies help limit exposure to chemicals and fumes.

Fiber End-Face Polish Techniques

The polishing technique used on fiber optic connectors depends on the connector ferrule. The fiber end face at the ferrule end may be finished in one of three ways: flat, PC-domed, or Angled-PC (Figure 6-4). The flat finish is accomplished by polishing the connector ferrule end on a glass (or hard plastic) surface. This finish produces a somewhat higher back reflection than other methods but is nonetheless acceptable for most multimode applications.

The most common of all finishes is the *domed or* PC type. In this case the polishing takes place on a rubber pad. This allows the fiber end face to become slightly rounded providing for contact of the cores only when the fibers are mated together in a mating sleeve.

Angled PC singlemode connectors are relatively new to the fiber market and use an 8-degree chamfer on the end face of the connector ferrule. These connectors produce the least loss and lowest back reflection of the three finish methods. They are, however, difficult to field terminate and cannot be mixed with either of the other two finish types.

Cleaning Fiber Optic Connectors

With fiber optics, tolerance to dirt is near zero. Airborne particles are about the size of the core of SM fiber and are usually silica based. They may scratch PC connectors if not removed! With most network cable plants, every connection should be cleaned during installation and not removed except for testing. Test equipment that has fiber-bulkhead outputs and test cables needs periodic cleaning, since there may be hundreds of insertions in a short timeframe. Here's a summary of what we have learned about cleaning fiber optic connectors.

1. Always keep dust caps on connectors, bulkhead splices, patch panels, or anything else that is going to have a connection made with it.

2. Use lint-free pads and isopropyl alcohol to clean the connectors. Some solvents *might* attack epoxy, so only alcohol should be used. Cotton swabs and cloth leave threads behind. Some optical cleaners leave residues. Residues usually attract dirt and make it stick. For over 10 years we have been supplying "Alco Pads" with every Fotec Test Kit with no problems.

3. All "canned air" now has a liquid propellant. Years ago, you could buy a can of plain dry nitrogen to blow things out with, but no longer. Today's aerosol cleaners use non-CFC propellant and will leave a residue unless you hold them perfectly level when spraying, and spray for three to five seconds before using to insure that any liquid propellant is expelled from the nozzle. These cans can be used to blow dust out of bulkheads with a connector in the other side or an active device mount (xmit/rcvr). NEVER use compressed air from a hose (This emits a fine spray of oil from the compressor!) or blow on connectors (Your breath is full of moisture, not to mention all those yucky germs!).

4. A better way to clean these bulkheads is to remove both connectors and clean with Alco Pads, then use a swab made of the same material with alcohol on it to clean out the bulkhead.

5. Detectors on fiber optics power meters should also be cleaned with the Alco Pads occasionally to remove dirt. Take the connector adapter off and wipe the surface, then air dry.

6. Ferrules on the connectors/cables used for testing will get dirty because they scrape off the material of the alignment sleeve in the splice bushing. Some of these sleeves are molded glass-filled thermoplastic and sold for multimode applications. These will give you a dirty connector ferrule in 10 insertions. You can see the front edge of the connector ferrule getting black. The alignment sleeve will build up an internal ledge and create a gap between the mating ferrules—Voila! A 1–2 dB attenuator! Use the metal or ceramic alignment sleeve bulkheads only if you are expecting repeated insertions. Cleaning the above requires aggressive scrubbing on the ferrules with the Alco Pad and tossing the bulkhead away.

7. Some companies sell a cleaning kit for fiber optics. These are good solutions but perhaps not as cost-effective as making your own to meet your needs.

SPLICES

Splices normally are a permanent joint between two fibers. The two basic categories of splices are fusion and mechanical. Generally speaking, splices offer a lower return loss, lower attenuation, and greater physical strength than connectors. Also, splices are usually less expensive per splice (or per joint) than connectors, require less labor, constitute a smaller joint for inclusion into splice closures, offer a better hermetic seal, and allow either individual or mass splicing.

Fusion Splicing

Fusion splicing (Figure 6-9) works on the principle of an electric arc ionizing the space between the prepared fibers to eliminate air and to heat the fibers to proper temperature (2,000°F). The fiber is then fed in as a semiliquid and melds into its mate. The previously removed plastic coating is replaced with a plastic sleeve or other protective device. The perfect fusion splice results in a single fiber rather than two fibers having been joined. One drawback to fusion splicing is that it most generally must be performed in a controlled environment, that is, a splicing van or trailer, and should not be done in open spaces because of dust and other contamination. Fusion splicing in manholes is prohibited because the electric arc generated during this process may cause explosions if gas is present. Due to the welding process, it is sometimes necessary to modify the fusion parameters to suit particular types of fibers, especially if it is necessary to fuse two different fibers (from two different manufacturers or fibers with different core/cladding structures).

Fusion splicers can be purchased rather plain, such as the fixed V-groove type, for as little as $10,000, or completely automated and capable of fusing 12 fibers in a ribbon simultaneously. To assure consistent low-loss splices, an automated

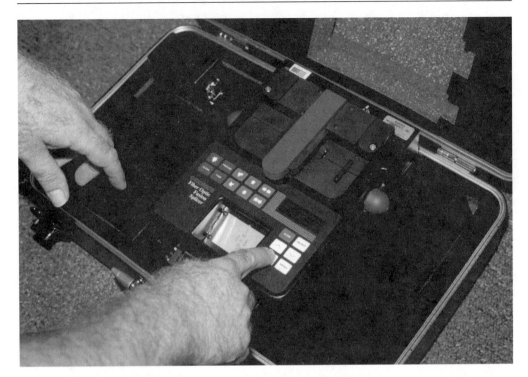

Figure 6-9   Fusion splicer.

splicer costing $25,000–40,000 can be acquired with features such as self-alignment and automatic loss testing.

Mechanical Splicing

Mechanical splicing, on the other hand, is quick and easy for restoration, its major use, and is also used for new construction, especially with multimode fiber. It does not require a controlled environment other than common sense dust control. The strength of a mechanical splice is better than most connectors; however, fusion remains the strongest method of splicing. Back reflection and loss vary dramatically from one type of splice to another. Equipment investment for specific splicing kits need not exceed $5,000.

Mechanical splices (Figure 6-10) employ either a V-groove or tube-type design to obtain fiber alignment. The V-groove is probably the oldest and is still the most popular method, especially for multifiber splicing of ribbon cable. Examples of this type include the 3M Fibrlok, Siecor CamSplice, AMP CoreLink, and the Lucent CSL splice.

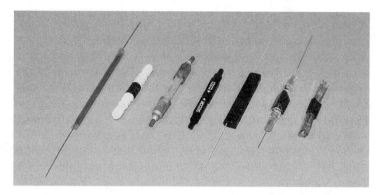

Figure 6-10   Typical mechanical splices and a fusion splice on the far left.

Tubular splices, on the other hand, require that the fibers be inserted into a small tube, which provides alignment. The splice is then glued or crimped to hold the fibers together. Examples of this type of splice include the Fastomeric, Elastomeric, the AMP Optimate, and the Norland Optical Splice.

Splices originally were either glued (GTE Elastomeric) or polished (AT&T Rotary). However, splices using adhesives have been phased out of common use because of the reusability of the more modem "no-glue" type. Polished splices are very much like miniaturized connectors using ferrules and a polishing process. Because of the extensive time required, polish type splices are also extinct.

In nearly all of today's splices, the fibers are *crimped* or *locked* to achieve fiber alignment and attachment. Most of them may be reopened for fine-tuning and possible reuse. These mechanical splices must all use some type of index-matching gel inside to eliminate back reflection and reduce splice loss. This gel is subject to contamination, so care is required when handling the splice, particularly if it is going to be reused later.

To prepare for mechanical splicing, the fibers are first stripped of all primary coating material, cleaned with alcohol, and then cleaved as previously described. Completed splices, whether fusion or mechanical, are then placed into splicing trays designed to accommodate the particular type of splice in use. Splicing trays then fit into splice organizers and in turn into a splice closure.

Choosing a Splice Type

The type of splice chosen is usually determined by the following criteria:

I. **Type of fiber:** Most singlemode fiber is fusion spliced because this results in lower loss and better return loss performance. Multimode fiber, with its complicated core structure, does not always fusion splice easily, so mechanical splices can give equal performance at a lower amortized cost.

2. **Attenuation, including return loss:** Today's automated fusion splicers can produce incredibly low-loss splices (typically 0.0 to 0.15 dB). Although a properly installed mechanical splice may also achieve a near-zero loss, the consistency of the fusion splice is hard to beat. The main difference between the two is the back reflection caused by the nature of mechanical splices.

3 . **Physical durability:** The welding process used in the fusion splice gives higher strength and greater durability. Fusion splicing retains the original mechanical tensile strength of the fiber, that is 50,000 to 75,000 psi. Most mechanical splices are rated at a pullout strength of no more than 1 to 2 pounds.

4. **Ease of installation:** A fully automated fusion splicer is very expensive but makes the splicing a one-button process. Mechanical splicing types vary but usually are less expensive to purchase and use for low-count fiber jobs.

5. **Cost per splice:** In the case of fusion splicing, which is the most common type of splicing being performed on singlemode fiber for new construction, the initial capital investment is much greater than the cost for mechanical splicing. A fusion splicing machine is a very large investment. Also, fusion splicing must be performed in a controlled environment, necessitating a splicing trailer or van. Mechanical splicing, on the other hand, requires no controlled environment, has a very low initial capital outlay, and the splices themselves vary from $7 to $20 each.

Terminating Singlemode Fibers with Pigtails

Singlemode cables are generally terminated using a combination of connector installation and splicing. Since singlemode connectors have such critical dimensions and mating surface requirements, they are generally terminated in a manufacturing lab. There the proper fiber insertion and physical contact polishing can be controlled precisely. Complete cable assemblies with connectors on both ends are made and tested, since testing a cable with two ends is easier than with bare fiber on one end. In the field, the assemblies are cut in half and spliced onto the installed backbone cables. Although the splice contributes some additional loss and cost, the overall method provides a higher yield and better connection at lower cost than trying to control the termination process in the field.

TOOLS

No job can be completed without the correct tools, and fiber splicing/connectorization is no exception. Following is a summary of the tools and test equipment needed.

Handtools

Handtools can be purchased in a prepackaged tool kit or on an individual basis as needed. At a minimum the following will be needed to complete most fiber optic operations.

1. Cleaning fluid and lint-free wipes (approved cable cleaner)
2. Buffer tube cutter
3. Reagent grade isopropyl alcohol (99%) in nonspill bottle or presoaked pads
4. Canned "air"
5. Tape: masking and Scotch invisible
6. Coating stripper
7. Cleaver or scribe
8. Microscope or cleave checker
9. Splicing method—fusion or mechanical—determines specific tooling needs
10. Connectorization method determines specific tool kit (if required)

The total cost of these tools can vary from as little as $750 to as much as $5,000 depending on the quality and quantity of tools purchased.

Major Equipment

1. Fusion splicer (optional)
2. Optical time domain reflectometer (OTDR) (optional, rentable)
3. Splicing van or trailer (nice to have an organized workplace, especially for outside plant work)
4. Power meter (for measuring optical power or loss)
5. LED or laser light source (to inject a test signal for loss)
6. Visible light source (for tracing cables, absolutely mandatory!)
7. Fiber optic talkset (to communicate over the fiber; alternatives are walkie-talkies and cellular phones)
8. Termination kits (these may be made by purchasing tools individually rather than in a kit form. Sometimes the splicing or connectorization kits will contain too many small insignificant tools that you may already own. Once you determine the needed tools, you can purchase only those tools.)

Since the major tools and test equipment represent a large capital investment, you should consider leasing the more expensive of these to determine which will fit your intended purposes. After a period of use, a purchase might be considered.

REVIEW QUESTIONS

1. Fiber optic joints should have _____
   a. back reflection.
   b. an index matching gel.
   c. high mechanical strength and low loss.
   d. attenuation.

2. Loss of light power is called _____
   a. dBm.
   b. attenuation.
   c. absorption.
   d. diffusion.

3. Identify the following as either an
   a. intrinsic loss.
   b. extrinsic loss.
   ____ core eccentricity   ____ NA mismatch
   ____ misalignment   ____ contamination
   ____ poor cleave   ____ core diameter mismatch
   ____ core ellipticity   ____ air gap between fiber ends

4. _____ connectors do not need to be polished.
   a. Epoxy glue
   b. "Hot Melt"
   c. Anaerobic
   d. Cleave and crimp

5. Match the terms right column with appropriate terms in left column.
   ____ Mechanical splicing   a. nonpermanent connections
   ____ Fusion splicing   b. permanent singlemode connection
   ____ Connector   c. restoration

7

FIBER OPTIC HARDWARE

STEVE PAULOV

The main purpose for hardware is to protect and organize splice and termination points. Hardware can be divided into two categories; indoor and outdoor. We deal primarily with indoor hardware in this chapter, as applicable to commercial buildings and campus environments, with a final review of outside plant hardware at the end of the chapter.

Fiber optic cable plant design requires preplanning to select the proper hardware. You need to know the number of fibers to pull to each location and the purpose for each fiber. From there you can plan your hardware and the type of cable to be used for riser, plenum, or other applications. You should also plan the conduit, duct, and inner duct. The designer should decide in advance on the networks, support systems, and topologies to be used, such as Ethernet, token ring, voice, data, video, imaging, control, and industrial applications. Once the application is determined and the fiber and connection/splice is known, the route must be determined. Then we can decide on locations for the hardware cabinets, racks, and associated panels.

There are rules for fiber distribution established by the EIA/TIA 568 Committee. The EIA/TIA 568 Commercial Building Wiring standard is based on a hierarchical star for the backbone and a single star for horizontal distribution. Rules for backbone wiring include a 2,000-meter (6,557-ft.) maximum distance between the main cross-connect and the telecommunications closet, with a maximum of

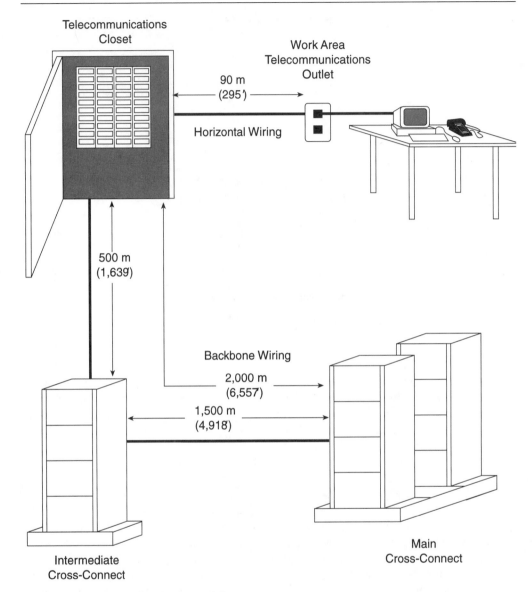

Figure 7-1 Backbone cabling.

one intermediate cross-connect between the main cross-connect and the telecommunications closet (Figure 7-1) Many networks are installed in a simpler, less expensive "homerun" design (Figure 7-2). The four main functions of each distribution point (using the EIA/TIA 568 nomenclature) are as follows:

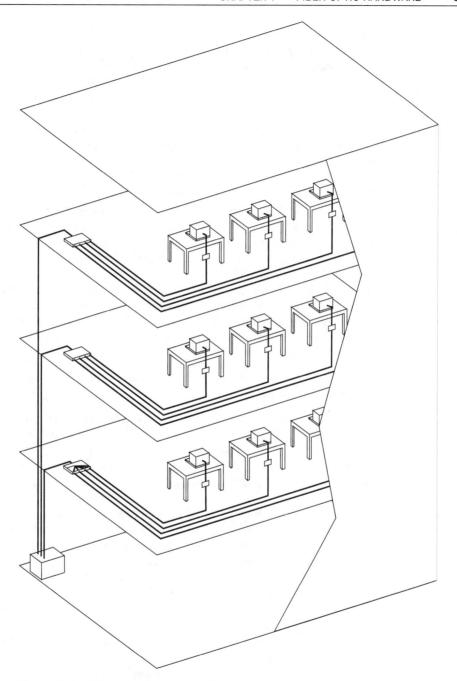

Figure 7-2 "Homerun" or centralized fiber cabling system.

1. Main cross-connect. The main cross-connect (MC), also referred to as the main distribution frame (MDF), should be in proximity to or in the same location as the data center or PBX. This placement ensures a centralized management point for reconfiguration of the fiber optic network. Equipment in the MC should be capable of the following:
   - Handling adequately large fiber counts
   - Accepting either direct termination of equipment or pigtail splices and preterminated assemblies for splicing at the frame
   - Providing jumper storage and reconfiguration capabilities
   - Allowing for growth

   Typically, MC equipment can be installed in racks or wall cabinets.
2. Intermediate cross-connect. The intermediate cross-connect (IC) typically connects the intrabuilding cable plant to the interbuilding cable plant. It is smaller in scope with lower fiber counts than the MC. Products in the IC may need to be wall mounted. The size of the IC will determine which products will be used.
3. Telecom closet hardware. Telecom closet (TC) equipment makes the transition from the backbone to the horizontal cable plant.
4. Work area telecom outlet. The work area telecom outlet is the end point of the horizontal wiring. The cables may be fiber only or a combination of fiber and copper.

In each of the above distribution points, one or more of the following hardware components can be installed.

1. Splice panel. The splice panel changes outside cable to riser or plenum cable or breaks out the outside cable to individual buffered fibers. After being spliced, the fibers are routed to the appropriate splice tray and positioned to prevent damage. The splice panel may accommodate fusion or mechanical splices and can be mounted in a rack or wall cabinet (Figure 7-3).
2. Patch panel. A patch panel provides a centralized location for patching fibers, testing, monitoring, and restoring riser or trunk cables. Figure 7-4 shows an incoming riser cable being terminated on one side of a coupling panel with a connector. On the other side of the coupling panel is another group of connectors terminating the LAN cables. In small offices, the patch panel can be used as an MC or IC. To order this type of panel, you must know the type of connectors for the coupling panels, the number of connections needed, the type of cabinet, whether the panel will be wall or rack mounted, and the size of the cabinet required for the number of panels to be installed.
3. Wall outlet. A wall outlet (Figure 7-5) terminates the permanent wiring and provides a connection for a short jumper cable to connect the network

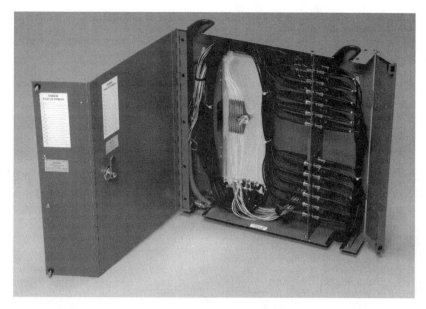

Figure 7-3   Splice panels connect individual fibers from cables to pigtails. Courtesy AMP

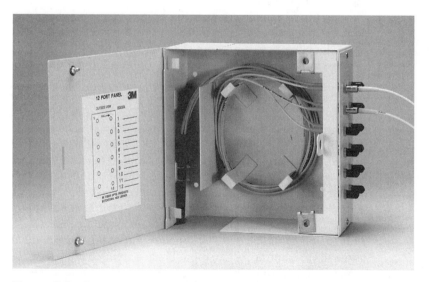

Figure 7-4   The patch panel makes connection between two cables. Here we see single fiber jumpers connected to a breakout cable. Courtesy 3M

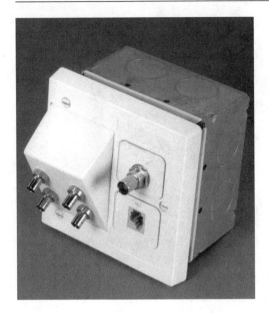

Figure 7-5 Wall outlets terminate cables in convenient locations for desktop connections. Courtesy AMP

equipment. Wall outlets can be wall mounted or permanently installed into the wall surface.

4. Panels. When designing your cabinets, you can select the panel with the type of coupling to fit the connectors you have chosen. You can also order coupling panels without the couplers and install these yourself if the type of connectors to be used will not be known until the cut-over date.

Each manufacturer has a unique design for stacking equipment panels. One such method involves stacking the panels on a relay rack. Another method places the panels in a cabinet, and each cabinet is installed on the wall above each other or side by side. When ordering any cabinet make sure it is equipped with cable-terminating equipment such as ground clamps, a place to secure the dielectric central member, Kevlar or other fill material, and cable ring or channel guides.

5. Pigtail splicing. A fiber optic pigtail splice is a fiber cable that has been factory connectorized on one end with an optical connector. The other end remains unterminated. The unterminated end of the pigtail is spliced to the fiber requiring termination. Pigtails are used almost exclusively for singlemode applications but occasionally for multimode. Generally pigtails are bought as terminated jumpers and cut in half for splicing. That way, they can be tested more easily to ensure the connectors are good before splicing them onto the fibers.

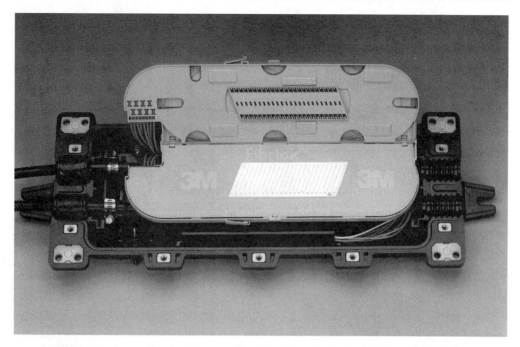

Figure 7-6   Outside plant splice closure for 3M Fiberlok mechanical splice.
Courtesy 3M

6. Outside plant hardware. For cable runs outside, one generally is concerned with splice closures (Figure 7-6), since connectors are used only at the terminated ends inside buildings. Splice closures are available for direct burial, use in controlled environment vaults (CEVs), aerial or pedestal installation. Most closures are sealed once splices are loaded. Closures must be chosen to be appropriate for the installation location and splice trays must fit the type of splice to be used.

7. Conduit. Fiber optic cable is often installed in conduit (Figure 7-7). The conduit can protect the cable from damage, water, or stress, both during and after installation. Conduit comes in many types from simple corrugated innerduct to steel pipe. Due to the concern over damaging fiber optic cable during installation, many types of conduit have been designed to facilitate installing long lengths of cable without exceeding the cable pull limits.

Even indoors, fiber is often installed in simple innerduct. This identifies the cable as fiber and protects it from damage during the installation or removal of other cables. It may also make installation faster, if the innerduct has preinstalled

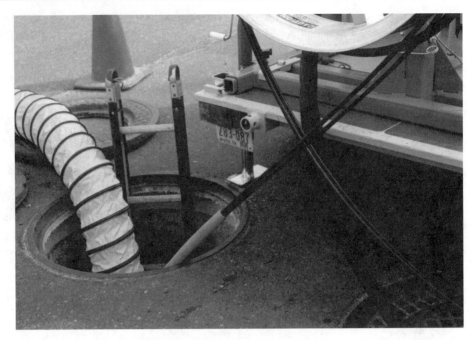

Figure 7-7   Two fiber optic cables being pulled through innerduct that has previously been pulled into conduit. Courtesy Condux

pulling tape. The duct can be installed by relatively unskilled personnel, the fiber installed quickly without as much concern for damage, since it is protected by the innerduct.

Outdoors, conduit and innerduct, along with proper lubrication, can extend the pulling length considerably. They can be chosen from a number of styles and sizes to fit the number and sizes of cables to be installed, especially considering future expansion. Some manufacturers even offer conduit or innerduct with fiber optic cable supplied to the customer's specifications preinstalled and ready for direct burial or pulling into preinstalled conduit.

Trenching for conduit: In paved areas, the surface should be carefully cut to prevent unnecessarily excessive width at the top of the trench and thus reduce the amount of surface to be repaved. For economical operation, particularly where paving is involved, the trench width should be no greater than is needed to provide adequate working space. Generally, this dimension is controlled by the types of excavating equipment used. As a minimum, the trench must be 4 inches wider than the width of the conduit structure where backfill will be used, and 3 inches wider where concrete encasement will be used. Individual job specifications will dictate trench width. Grade and level the trench bed. Where necessary, provide sand and/or other granular backfill as bedding material so the conduit will be evenly supported over the length of each section.

In order to make optimum use of the conduit for subsequent cable placing operations, particular care should be taken concerning direction changes. The ideal duct structure is one that is essentially straight and allows for grading for drainage into manholes. Direction changes should be made as gradual as possible.

When a road bore has to be made, the recommended procedure is to install a pipe that is large enough for the joined sections of conduit to slide completely through. In an instance where this is not possible and only a 4-inch steel pipe is present for the bore, the innerduct can be used for insertion through the steel pipe.

REVIEW QUESTIONS

1. _____ provides a centralized management point for network configuration.
   a. Work area telecom outlet
   b. Telecom closet
   c. Intermediate cross-connect
   d. Main cross-connect

2. _____ make the transition from the backbone to the horizontal cable plant.
   a. Work area telecom outlet
   b. Telecom closet
   c. Intermediate cross-connect
   d. Main cross-connect

3. The main purpose of hardware is to _____
   a. splice fibers.
   b. protect and maintain fiber optic cable.
   c. simplify pulling cable.
   d. provide an airtight enclosure.

4. The ideal duct structure is _____
   a. indoors.
   b. straight and allows for drainage.
   c. color coded.
   d. 4 inches wider than the conduit.

8

GETTING STARTED IN FIBER OPTICS

PAUL ROSENBERG

Contractors have many reasons to be interested in the installation of fiber optics. Since contractors already know how to install cables, the work is generally clean, and it is easier to supervise than most electrical projects. However, you need to know more about fiber optics than installation techniques to have a successful fiber optics business. Learning about the basics of fiber optics and getting hands-on training in the techniques of installation are important and essential subjects, but in reality, they are only the first step to a business in fiber optics. The problem is that getting started in fiber optic installations is not without pain.

The first obstacle you have to deal with is an investment in training. Training can be acquired in several ways, some of which can be expensive. No matter how the training is handled, it must be done well. It would be business suicide to send an untrained crew out to terminate fibers. The cost of replacing just one cable run would cover a significant amount of training expense. And any way you look at it, teaching someone to perform fiber optics installation is not as simple as teaching someone to bend 1/2-inch thinwall.

One training method is to send your people to a special school. Although this usually provides excellent training, it can be expensive, and usually your people will get hands-on experience with only one brand of equipment. Alternatively, you may send your people to an installers' conference to give them hands-on experience with products from a variety of manufacturers.

You may wish to train one or two of your best people and have them teach the rest of the crew. Make sure the ones you send are very good teachers. When they return, you must give them plenty of time to teach their newfound skills to the others. Make sure you buy enough materials for them to practice with (and throw away when they are done).

Another obstacle to getting into the fiber optics business is the cost of tools and equipment. This cost is not necessarily a tremendous impediment to getting started in fiber optics; it is one that continues to grow as a fiber optic business grows. For example, it is not difficult to get started installing multimode fiber, glued terminations, and perhaps mechanical splices. You need only a termination kit for the type of connectors you will install and some basic test equipment.

It is more difficult (and more expensive) if you want to take on a singlemode project. You will probably need to buy or rent a fusion splicer and new types of ancillary equipment. Later, if you want to take a larger project, you will need a better fusion splicer, and so on. And due to the quick pace of technology, your tools and testers may become obsolete long before they are worn out.

The difficulty of having to build business relationships with an entirely new group of people is another obstacle to getting into fiber optics. Electrical contractors have been dealing with electrical wholesalers for many years and are quite used to the process. Unfortunately, electrical wholesalers do not always deal with fiber cables and associated materials, although the number of those involved with fiber optics is growing rapidly. This means electrical contractors are forced to deal with new people, buying directly from manufacturers, through a specialty supply house, or via mail order.

Another obstacle for contractors is keeping up to date with technologies. Since today most technologies change quickly, information has a short life span—maybe three or four years. That means that if you were to learn everything possible about fiber this week, the information could be almost entirely out of date three or four years from now. Resign yourself to this idea. Subscribe to the appropriate magazines and devote some time regularly to reading articles about the state of the art technology.

WADING INTO THE MARKET

The best way to enter the market as an installer of optical fiber is to wade in slowly.

1. Learn as much about the technology as possible. Learn about the fibers, the cable types, testing, and terminations. Also learn about communications and data networks.
2. Train your workers in installation and termination techniques. Buy some basic tools and test equipment and give them time to practice. Make sure that they are competent before you go on to the next step.

3. Contract a small job. Take one small job and give it lots of attention. Watch all of the details on the project and note any difficult or unexpected parts. You can probably either subcontract a small installation from another electrical contractor or contract directly from one of your existing customers.

4. Begin to advertise. (We go through more details on advertising in a moment.)

5. Step up to larger and more complex projects. Move up slowly, and never take more than one step at a time. For instance, moving from a 50-termination project to an 80-termination project or from a 50-termination multimode to a 50-termination singlemode project is fine. However, going from a 50-termination multimode project to an 80-termination singlemode project may not be a good idea.

Remember, initially you will sell installation of cables, termination of fibers, splicing, and testing to customers. As you proceed further into the fiber optics business, you can sell design services as well.

GETTING BUSINESS

The best place to get fiber optic installation business is from your existing customers, although how much fiber work you can get from them will vary depending on their businesses. One place to get business is from other contractors who need a subcontractor with fiber experience.

Some other typical fiber customers are process-control people; network managers in business, college, and institutional settings; utility companies and cable television companies. Also work on getting referrals from distributors and component dealers. These people often have customers who ask them about installers. Give your vendors a sales pitch and ask them for their referrals.

Advertising is a good way to get business. The first thing to do is to put together a brochure. Be sure the brochure stresses experience and references. To get started in fiber optics you must convince customers that you know what you are doing. Distribute the brochure to your existing customers. Next, you should send copies of the brochure to all potential customers in your area. Follow up on the mailing with telephone calls to the recipients, and follow up on the calls with another mailing. Remember that ad campaigns such as this take time, but they are usually fairly successful.

As an installer of optical fiber systems, your customers will expect you to act like a consultant. This means you will need to dress appropriately for customer visits, and in general behave more like a consultant. (Of course, this means you can charge more for your design and troubleshooting services, just as does a consultant.)

Customers will expect you to understand communications technology and to help them find answers to their communications problems. Also, you will have to understand data networking, since a big part of your job will be helping your customers switch from copper to fiber networks and control wiring.

Tooling Up

Certain items are absolutely necessary for fiber work. If you are going to get started in the fiber optics business, you must have the right equipment to get the fibers in place and terminate and test them. No matter how you start, you will need the following:

- A test kit or power meter and source for loss testing
- Swivel pulling eyes
- Breakaway swivel
- Microscopes for inspecting polished connectors
- Termination kit or polishing tools, polishing films, adhesive syringes, cleavers, stripping tools, solvent and wipes, canned air, and adhesives

These things are in addition to the equipment and tools you already have for the installation of wire and cable. Fortunately, most of the items on this list are relatively inexpensive. In general, it is best to rent an OTDR if you need one for a singlemode fiber project. They generally are not needed for short-length multimode projects. The price of OTDRs is too high unless you use them all the time, and prices should continue to drop. It does not make sense to put out a significant amount of money for equipment until you get firmly entrenched in the optical fiber business.

OVERSEEING FIBER INSTALLATIONS

With fiber work, getting a job right is more difficult than getting it done on time. This is why it is critical that you know how to supervise the work of your people. Be willing to look over their shoulders on the job to make sure they are doing things right

Mistakes in fiber work, especially regarding terminations, are difficult to detect. A mistake that could keep the entire system from working might not show up until the system is completely installed and turned on. (Think about that for a minute—it is a scary situation.)

Job number one is to assure that all the terminations are done correctly. Terminating fiber is delicate work; make sure that your people work like jewelers, not like framers.

Make sure your people have all the right parts; make sure they don't rush; make sure they have a well-lit work area; make sure they have test equipment; and make sure they use it!

Next, make certain they mark every run of cable and termination well. Invest money in cable markers and numbers; invest time in written cable and termination schedules and cable plant documentation—DO NOT lose track of which cable is which.

Inspection

Inspection is a bit of a wild card in fiber installations. Since optical fibers do not carry electricity, they are not exactly the domain of the electrical inspector. And since electrical inspectors do not always inspect communications wiring, they are even less likely to inspect fiber installations. Nonetheless, take a moment to check with local electrical inspectors before you do work in their jurisdiction. Also, make sure you are familiar with the requirements of National Electrical Code Article 770.

Always make sure that you know who will inspect your work before you give your customer a final price. You must know what the inspector will expect of you and what he or she will be looking for. In most cases, the inspector of a fiber installation will be the person who signs the contract. In some cases, it will be a third party. Be especially careful of third-party inspectors, since they are getting paid to find your mistakes.

The bottom line in installation quality is signal strength from one end of the network to the other. But be careful of other details that may be noticed by an inspector. Among other things, many inspectors will give a lot of attention to proper cable marking, mechanical protection, and workmanship.

HOW MUCH DO I CHARGE?

Once you have made the decision to enter the fiber installation market, you have to decide how to come up with prices for the jobs you will bid. You also will need to become familiar with the RFP process (RFP stands for "request for proposal"). In the network world, an RFP is more or less the same as a bid document, except it is not as detailed. An RFP gives you the general details of a project and asks you to furnish a complete design, schedule, and price. Completing the RFP process is similar to figuring and bidding a design/build project.

The first thing you have to do when putting together an estimate is figure out your real costs. Add an appropriate amount for overhead to these costs. This should leave you with a comfortable base price. (If not, raise the price until you are comfortable. Never intentionally take a job that will cause you to lose money.)

Pricing fiber optic work is difficult because it is easy to leave things out. To avoid such errors, think through the design of the installation carefully. Also, put a bit of contingency money in your estimates to compensate for surprises. A small percentage of your total cost should be all that you need. This is especially important while technology continues to change so quickly.

In general, the parts of an optical fiber installation that you must account for are cables, terminations, splices, jumpers, attenuators, breakout kits, distribution cabinets, cross-connects, patch panels, outlets and jacks, grounding clamps, cable lubricants, pulling hardware, conduit, inner duct, racks, tie-wraps, and cable markers.

For pricing very small fiber installations, it is generally best to calculate your material expenses first, and then calculate your time in bulk. That is, figure your labor as X number of days for a two-man crew. On larger jobs, you should use the itemized estimating method you use for standard power estimates. You can find fiber labor units from different industry sources. Developing your own is a pretty big chore.

General guideline numbers can be useful when putting together an estimate. Remember, however, that such numbers are "for informational purposes only." Do not use them for pricing your own projects; it is the height of foolishness to take a contract without knowing your costs.

Once you get competitive with fiber bidding, remember that estimating and bidding are very different operations. The purpose of estimating is to know the lowest price for which you can perform any given project and not lose money. The purpose of bidding is get the highest possible price for a project. For more detailed guidelines on estimating, see Chapter 14.

REVIEW QUESTIONS

1. The best place to get fiber optic installation business is from _____
   a. trade journals.
   b. existing customers.
   c. vendors.
   d. distributors.

2. When first starting out in fiber optics, if needed, it is best to rent a(n) _____ rather than purchase one.
   a. power meter
   b. termination kit
   c. OTDR
   d. pulling eye

3. The purpose of a(n) _____ is to get the highest possible price for a project.
   a. RFP
   b. bid
   c. estimate
   d. guideline

4. The purpose of a(n) _____ is to know the lowest price to complete a project and not lose money.
   a. RFP
   b. bid
   c. estimate
   d. guideline

3. Installed fiber must be _____
   a. documented.
   b. tested.
   c. terminated correctly.
   d. all of the above.

9

GUIDELINES FOR FIBER OPTIC DESIGN AND INSTALLATION

ERIC PEARSON

GENERAL GUIDELINES

Use these guidelines for evaluating and comparing vendors, choosing products, and evaluating options and cost impacts of your decisions.

1. **Start with a list of specifications.** After the cable plant is designed, but before you contact any manufacturers, copy the information you have generated onto a specification summary sheet so you have all of the specifications for each component conveniently available when talking with potential suppliers.

2. **All performance numbers need test specifications.** In addition to learning whether the numbers are minimums, maximums, or typical performances, you need to know how the numbers were determined. When comparing products from different vendors, be certain that the specifications are the same, so that you know that you are comparing apples to apples.

3. **Talk to at least four vendors for each component.** Remember that the fiber optic industry is highly competitive. You will be able to ask for, and

usually receive, the best price possible, even though you may be comparing several products that are not identical, such as cables or connectors. Remember to ask the question "Why should I buy your product instead of another supplier's product?" The answer to this question may include mention of other suppliers or of equivalent products.

4. **Talk with representatives more knowledgeable than yourself—avoid the "blind leading the blind."** By this step in the process of designing your fiber optic system, you should be more knowledgeable than many of the individuals selling fiber optic products. Therefore, when you need information, deal with people who know what they are talking about, even if you have to bypass the salesmen or representatives and talk directly with the factory.

5. **Expect and resolve conflicting facts/opinions.** At this step in the process of designing your fiber optic system, you have defined the performance criteria that will result in the lowest-cost fiber optic system possible. Now, you need to find sources for the components for which you have specifications. Be aware that you will probably learn from potential suppliers some facts that differ from the facts and information you may have learned elsewhere. Expect this to happen. When it does, you should resolve these differences.

6. **Use the five-year rule.** Unless you have an excellent and compelling reason not to do so, deal with suppliers who have been in business for at least five years. Such companies have been in business long enough to know how to avoid most of their problems, how to guide you to avoid your problems, and how to be around in another five years (when you need repair or some other support).

7. **Use the "high serial number" rule.** Unless you have an excellent and compelling reason to do so, avoid buying a product with a low serial number [or kilometer number]. For any product other than the most simple, you are less likely to experience problems if the serial number is 100 than if it is 11. This and the previous guideline should not limit the number of vendors who can supply what you need. When talking with potential suppliers, ask questions such as How long have you been making this product? How many customer returns or field problems have you had with this product? and How many kilometers have you made of this product? The answers to these types of questions will help you evaluate the reliability, workmanship, quality, and delivery considerations that you will use in deciding on a supplier.

8. **Know the cost impact of your design decisions.** Know or learn the cost impact of your choice of one design over another. This guideline applies equally to optoelectronics (wavelength of operation and light source-LED

or laser diode), to connectors (ST versus ESCON or FDDI), and to cables (loose tube versus breakout distribution).

9. **Buy only the performance you need.** Do not be influenced by an answer that includes any phrase such as "our product has higher or better performance that justifies its higher price" unless you need that "better" performance. In general, most products available in today's marketplace exceed the performance needs, either in distance of transmission, or of bandwidth, or of accuracy of transmission, of most applications. You will gain nothing by paying a premium price for "higher" performance that you will not use.

10. **Ask for competitive information.** Ask questions such as Who is your competition? Or Who do you compete against? You may not get an answer, but sometimes you will find new suppliers this way.

CABLE GUIDELINES

1. **Ask for product close to your needs.** After you have presented your list of specifications, ask whether each supplier has any other products available at a lower price. Cable manufacturers will often have leftover lengths, lengths with a single broken fiber (for example, 1 broken fiber out of 12 when you need 10), or lengths with one fiber with an attenuation rate higher than the specification for which it was originally manufactured. These types of products may well suit your needs. They can often be purchased at a significant savings over the "exact" product you need, if they meet your needs without compromising performance.

2. **Look for the best price possible.** The fiber optic cable industry is highly competitive, so you will be able to get a good price by shopping around. It has been our experience that at any point in time, there is always some manufacturer hungry enough for business to offer exceptionally low prices. In some cases, manufacturers have been selling cable at a loss in order to capture market share. You should contact at least four suppliers, but should not need to contact more than seven or eight in order to get the best price possible.

3. **Be cautious about premium prices.** Be cautious about paying premium price for premium performance, unless your design calculations support this need. Most data communication applications do not use the optical power budget available or the bandwidth distance product. Before paying premium prices for premium bandwidth-distance product (BWDP), be certain to perform the effective bandwidth calculations. These calculations will demonstrate that 850-nm operation usually does not justify premium BWDP. These calculations may justify premium BWDP for

1300-nm operation. And you can safely assume the fiber being used all exceeds manufacturers' minimum specifications by a large margin.

OTHER CONSIDERATIONS

Of course, price is not the only consideration. Reliability, workmanship, quality, and delivery are also important. In order to evaluate these other considerations, we provide the following observations. We have found that most manufacturers of fiber optic cable have about the same level of reliability and other considerations. In addition, as we review the data sheets of comparable cables, we find that no supplier has the *best* products. Finally, our conversations with both purchasers of cables and others in our industry support this observation: Most manufacturers are reasonably reliable in the products they supply to their customers. This is as it should be, since most cable manufacturers have been in business for at least five years.

CONNECTOR GUIDELINES

1. **Are any connectors easier to install?** Remember to be skeptical of any salesperson who tells you that his or her product has the "best performance," "highest convenience," or "easiest installation." Most of these statements are sales propaganda. The most accurate statement is that some connector installation techniques are easier than others. The reality is that there is no universally "best product." All fiber optic connector types have advantages and disadvantages, which will add up differently depending on the factors in your specific situation. These factors are many: product performance, product price, cost of installation tools and fixtures, the number of connectors being installed, the level of experience of the installer, the need—or lack thereof—for training, and a number of additional factors.

2. **Know about ceramic connectors.** When considering connectors with ceramic ferrules, remember that most ceramic ferrules are manufactured by the same few Japanese companies. As such, there should be no significant price differences from different manufacturers.

3. **Ease of installation may result in undesirable performance.** Some connectors are sold as "requiring no adhesive or epoxy." The sales pitch here is ease of installation. To be certain, the use of an adhesive or epoxy is time-consuming and inconvenient. However, these other methods may result in other performance disadvantages. For instance, all of the connectors not requiring these materials use a clamping mechanism for gripping the fiber. Use of this mechanism can allow the fiber to move into

and out of the connector, called "pistoning." In some situations, this pistoning will not result in any problems. However, in other situations, it will cause severe problems, including destruction of active devices in the optoelectronic boxes. The point in both of these examples is that your choice of connectors needs to be made on the basis of performance and cost, not solely on the basis of style or design, and not solely on the basis of any single characteristic.

4. **There is no substitute for experience.** Whatever components you choose to generally use, you must learn how to use them and practice doing it continuously to not lose the technique. This is especially important for connector termination and splicing. Unless you are doing it continuously, you must work your way back up the learning curve each time you begin a new job. Your yield (number of good connectors out of the number attempted) will be a function of how much practice you have. Setting aside some time to practice before going on a field installation may be an excellent investment.

REVIEW QUESTIONS

1. What is the first thing needed to start evaluating fiber optic vendors, products, and pricing?
   a. How long the manufacturer has been in business.
   b. A specification summary sheet for each component.
   c. Alternative product options.
   d. Experience with the product or vendor.

2. What is the most important thing to determine when talking to a representative of a fiber optic manufacturer?
   a. How long the person has worked for the company.
   b. Who is the company's competition.
   c. Is the person knowledgeable about the products and applications such as yours.
   d. all of the above

3. Which of the following are true about these connector specifications? Ease of installation, High performance, Low cost
   a. T or F: Ceramic ferrule connectors have about the same performance regardless of cost, so you should buy for price and delivery.
   b. T or F: High performance connectors are easier to install.
   c. T or F: Connectors that are easy to install are more expensive to buy but may not be more expensive when labor is included in the cost.

10

CABLE PLANT LINK LOSS BUDGET ANALYSIS

DAVE CHANEY

In order to operate properly, a fiber optic network link must have an adequate loss margin. That is, the total loss in the installed cable plant must be less than the tolerable loss of the transmitters and receivers in the transmission equipment being used. Figure 10-1 graphically illustrates the link loss parameters.

During the design phase, the cable plant loss must be estimated, based on average component specifications and the total cable length, to ensure the chosen equipment will work properly. Ideally, there should be at least 3 dB less loss in the cable plant than the link dynamic range to allow for component degradation and potential restoration splicing.

Loss budget analysis calculation and verification of a fiber optic system's operating characteristics includes all items in the cable plant, such as fiber length, number of connectors and splices, and any other passive components such as optical splitters. Optical loss is the key parameter for loss budget analysis, but bandwidth must be considered in some high bit-rate multimode systems such as FDDI, where a maximum cable length is specified regardless of optical loss.

Prior to implementing or designing a fiber optic circuit, a cable plant loss analysis is required. Prior to system turn up, test the circuit with a source and fiber optic power meter to ensure that it is within the loss budget.

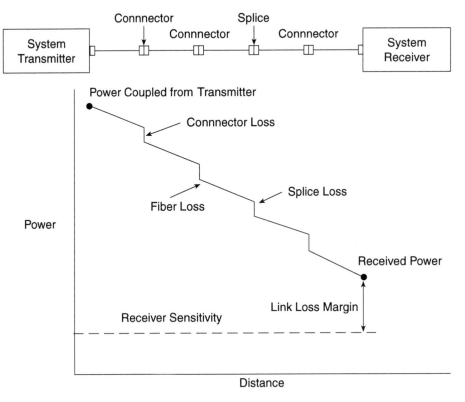

Figure 10-1   Fiber optic link loss budget.

CABLE PLANT PASSIVE COMPONENT LOSS

Consider the link shown in Figure 10-1 for operation at 1300 nm on multimode fiber.

Step 1. Fiber Loss at the Operating Wavelength

Cable length (km)	2.0			

	Multimode		Singlemode	
Typical fiber loss:				
Fiber type	Multimode		Singlemode	
Wavelength (nm)	850	1300	1300	1550
Fiber attenuation (dB/km)	3	1	0.5	0.4
Total fiber loss for system operating on multimode fiber at 1300 mn		2.0		

Step 2. Connector Loss

Typical connector loss	0.5 dB
Total number of connector pairs	5 (including connectors on ends of cable.)
Total connector loss	2.5 dB

Step 3. Splice Loss

Typical splice loss	0.2 dB
Total number of splices	1
Total splice loss	0.2 dB

Step 4. Total Cable Plant Attenuation

Total fiber loss (dB)	2.0
Total connector loss (dB)	2.5
Total splice loss (dB)	0.2
Other (dB)	0
Total link loss (dB)	4.7

EQUIPMENT LINK LOSS BUDGET CALCULATION

Step 5. From Manufacturer's Specification for Active Components

Operating wavelength (nm)	1300
Fiber type	MM
Receiver sensitivity (dBm@ specified BER)	−31
Average transmitter output (dBm)	−18
Dynamic range (dB)	13
Recommended excess margin (dB)	3
Maximum cable plant loss (dB)	10

Step 6. Loss Margin Calculation

Dynamic range (dB)	10
Cable plant link loss (dB)	−4.7
Link loss margin (dB)	5.3

This calculation must be verified by testing with a source of the proper wave length and a power meter after installation. In some cases, where the equipment may be operating on two different wavelengths or future upgrades are planned, testing at two wavelengths may be required. If the calculated and tested values differ considerably, but all the tested fibers in the cable are similar in loss, remember the specifications used for calculations are not exact. As long as proper operating margins are available, the cable plant should be acceptable.

REVIEW QUESTIONS

cable length: 3 km
number of connections: 2
number of splices: 1
operating wavelength: 1300 nm
fiber type: MM
receiver sensitivity: –35 dBm
average transmitter output: –25 dBm

Using this example fiber link and the typical losses on page 122, determine the following.

1. Connector loss: _____
   a. 1 dB
   b. .5 dB
   c. 2 dB
   d. 3 dB

2. Total cable plant loss: _____
   a. 7.2 dB
   b. 4.2 dB
   c. 5.2 dB
   d. 2.7 dB

3. Maximum allowable loss (with excess margin factored in): _____
   a. 10 dB
   b. 5 dB
   c. 7 dB
   d. 2.8 dB

4. Link loss margin: _____
   a. 3 dB
   b. 2 dB
   c. 1.8 dB
   d. 2.8 dB

5. Will the above fiber link loss be acceptable if the wavelength was changed to 850 nm? Assume the transmitter and receiver specifications remain the same.

   _____ Yes   _____ No

11

FIBER OPTIC
INSTALLATION SAFETY

PHIL SHECKLER

When speaking of safety in fiber optic installation, the first image that most people conjure up is a laser burning holes in metal or perhaps burning off warts. These images have no relevance to fiber optics. Optical sources used in fiber optics are of much lower power levels and are not focused into a time spot like these applications.

In fact, most datacom links use LEDs of very low power levels, and even the lasers in most fiber optic installations are of relatively low power. The light that exits an optical fiber is also spreading out in a cone, so the farther away from the end of the fiber you are, the lower the amount of power striking a given spot. Furthermore, the light is of a wavelength that cannot penetrate your eye because of the absorption of the water in your eyeball at those wavelengths. In order to do any damage, the end of the fiber would have to be held against your eyeball for hours! A complete safety study and report is in ANSI Z136.2, if you want to read the details.

BARE FIBER SAFETY

However, fiber optics installation is not without risks. As part of the termination and splicing process, you will be continually exposed to small scraps of bare fiber, cleaved off the ends of the fibers being terminated or spliced. These scraps are very dangerous. If they get into your eyes, they are very hard to flush out. The ends are extremely sharp and can easily penetrate your skin. They invariably break off and are very hard to find and remove. Most times, you have to wait for them to infect and work themselves out, which can be painful!

Wear safety glasses at all times when working with bare fibers. Be careful not to stick the broken ends into your fingers. Dispose of all scraps properly. You can keep a piece of double-stick tape on the bench to stick them to or put them in a properly marked paper cup or other container to dispose of later. Do not drop them on the floor where they will stick in carpets or on shoes and be carried elsewhere. And obviously do not eat anywhere near the work area.

OTHER CONSIDERATIONS

Fiber optic splicing and termination use various chemical cleaners and adhesives as part of the processes. Normal handling procedures for these substances should be observed. Even simple isopropyl alcohol used as a cleaner is flammable and should be handled carefully. Note that fusion splicers use an electric arc to make splices, so care must be taken to ensure no flammable gases are contained in the space where fusion splicing is done.

Smoking should not be allowed around fiber optic work. The ashes from smoking contribute to the dirt problems with fibers, in addition to the possible presence of combustible substances.

FIBER OPTIC INSTALLATION SAFETY RULES

1. Keep all food and beverages out of the work area. If fiber particles are ingested, they can cause internal hemorrhaging.
2. Wear disposable aprons to minimize fiber particles on your clothing. Fiber particles on your clothing can later get into food, drinks, and/or be ingested by other means.
3. Always wear protective gloves and safety glasses with side shields. Treat fiber optic splinters the same as you would glass splinters.
4. Never look directly into the end of fiber cables especially with a microscope until you are positive that there is no light source at the other end. Use a fiber optic power meter to make certain the fiber is dark. When using an optical tracer or continuity checker, look at the fiber from an angle at least 6 inches away from your eye to determine if the visible light is present.
5. Only work in well-ventilated areas.
6. Contact lens wearers must not handle their lenses until they have thoroughly washed their hands.

7. Do not touch your eyes while working with fiber optic systems until your hands have been thoroughly washed.
8. Keep all combustible materials safely away from the curing ovens.
9. Put all cut fiber pieces in a safe place.
10. Thoroughly clean your work area when you are done.
11. Do not smoke while working with fiber optic systems.

REVIEW QUESTIONS

1. The major safety concern regarding fiber optics is _____
   a. high power levels.
   b. warts.
   c. bare fiber ends.
   d. explosion.

2. Double-stick tape can be used for _____
   a. removing fiber splinters.
   b. protecting bare fiber from dust.
   c. collecting bare fiber ends.
   d. none of the above.

3. Fusion splicers should not be used _____
   a. with multimode fiber.
   b. with singlemode fiber.
   c. in a flammable space.
   d. near an open flame.

4. The most important installation safety rule is _____
   a. to work quickly when using epoxy or solvents.
   b. to brush any fiber ends off your clothes.
   c. to wear safety glasses.
   d. to work with a partner.

12

PLANNING THE
INSTALLATION

THOMAS A. DOOLEY AND
JERALD R. ROUNDS

THE PROCESS OF PLANNING

Planning is one of the most important parts of any project and is particularly critical in construction. One of the characteristics of construction is that it is not repetitive. Although specific activities such as pulling cable or making terminations appear to be repetitive, each is done under different circumstances. The cumulative result of these small differences is that jobs differ considerably. As a result, each job must be planned in detail to take its unique characteristics into consideration.

In manufacturing, prototypes are designed and manufactured to test the design and the manufacturing procedures. When mistakes and oversights occur, another prototype is developed to solve the problems. This process is repeated many times over until both the product and the process of making that product are flawless, at which time the product goes into production.

In the construction industry, you do not have that luxury. Designers must design correctly the first time. Constructors must build right the first time. Errors, either in design or construction, are costly to repair or replace and cause time delays. Some errors are hidden and do not become obvious until the system is operating. Then, the mistakes become not only costly and time-consuming, but are disastrous to those who depend on a fully operational and properly working system.

The only way to produce a quality job is to avoid errors, omissions, and mistakes by properly planning and by establishing correct installation procedures. A side benefit of the planning process, of course, is that it improves the efficiency of installation, which results in higher profits for a contractor. Preplanning of every aspect of fiber optic installation is not an option; it is a necessity.

MEASURING FOR CONDUIT PULLS

Planning for proper cable lengths is extremely important in fiber optic installation for two reasons. The most important is that splices cause loss in both signal quality and strength—this should be avoided. In fact, most designers specify point-to-point or device-to-device runs to avoid needless signal loss due to splices. The second reason, especially important to the contractor, is that splices cost time and money.

Measurements must be taken by an experienced field hand who understands the importance of correct and accurate measurements. Often, this is done by the project engineer or the job site superintendent.

There are three ways to measure the path to be used for the fiber. They are listed from least to most desirable.

1. Measurements taken from the set of prints. This works if you want to invest in a great deal of extra material. The objective of design drawings is to show generally where runs are supposed to be located. However, it is virtually impossible for the designer to anticipate all field conditions, and it is not the designer's intent to do so. Thus, taking accurate measurements from the design drawings is not possible and results in runs that are, at best, marginally long or short. Long pieces of fiber must be recut and short pieces are wasted.

2. Field measurements taken with a wheel. A site visit with a measuring wheel and a set of drawings will yield much better accuracy. Certain characteristics of the job site are given perspective with a field visit that are not obvious from inspection of the drawings alone. During the site visit, the drawings can be verified for accuracy, installation details such as pulling locations can be noted, changes such as variations in alignment or elevation can be identified, obstacles that might hinder the pull can be discovered, and termination locations can be recorded.

3. Measuring with measurement/pull tape. By far the most accurate and efficient method of measurement is with measurement/pull tape (sometimes called mule tape). This is a flat ribbon, consecutively numbered in feet, that is usually made with polyester or aramid yarn and sometimes

coated with plastic for waterproofing. It comes in different lengths and pull strengths.

Completing the measurement process requires adding extra length for splices, terminations, and future access coils, if specified. The amount of additional cable needed for splicing depends on site conditions, splicing method used, and long range plans for cable usage. Termination needs only about 7 to 10 feet of additional cable beyond the place the terminal will be mounted. Access or repair coils generally range in length from 30 to 50 feet per span. A good guideline is to allow 1 percent extra for outside plant cable and 5 to 7 percent extra for inside cable.

SPLICING

Fusion splicing cannot be done in explosive environments such as manholes. Therefore, each end coming into the splice that will finally end up in the manhole will have to have enough length to reach a tent or van near the manhole where the splice will actually be made. Usually, about 30 feet of extra length should be added for each side of the splice—resulting in additional 60 feet of cable.

Mechanical splicing can be done inside manholes. Mechanical splices require only about 10 feet of overlap, instead of the 60 feet for a fusion splice. Remember that these measurements are in addition to whatever length will be required to properly rack the cable within the manhole.

TERMINATIONS

When terminating fiber, it is important to place the terminal in a safe, noncongested area. If the end user has no preferences, place the terminal as close as possible to the fiber optic transmission equipment. It is most important to protect the exposed fibers, so choice of a termination location must consider a working environment that allows adequate working space.

EFFICIENT PULLING

When installers first become involved in fiber optic installations, they can easily become overconfident about how much cable length can be successfully pulled at one time. This is probably because of the small size and weight of fiber optic cable. It is better to divide a pull in half, or even thirds, usually at corners or pull boxes, than to fail in pulling a long run.

A pulling operation must be discontinued when the pulling tension reaches the cable limit. The discontinued pull will have to be aborted, the cable pulled back out and replaced on the reel, and the pull started again, either with shorter runs or better lubrication or both. The cable may be reused if the pull or extraction has not damaged it.

Every effort must be made to plan the pull so that pulling locations have plenty of work space. The knowledgeable designer will designate several appropriate locations for breaking up the pull. The installer must ensure that if the designer has not indicated where multiple pulls are necessary, he or she can determine appropriate locations for intermediate pulls before work is begun in the field. It is better to tie up one lead technician before the pulling starts than the whole crew after it has begun.

Most point-to-point pulls can be accomplished by crews of two or three people. One person pays the cable off the reel and into the duct in order to reduce tail load (the term used to designate that force required to get cable off the reel and into the duct). One or two people pull at the other end. The size of crew is determined by considerations such as:

- Length of pull
- Total degrees of bend
- Tail loading
- Use of lubricant
- Use of power pulling equipment

When planning to start a pull, the installer must make sure there is enough time to finish the job the same day. It is unwise to leave half a reel of cable, worth anywhere from $5,000 to $375,000, lying around in unsecured areas. If a reel must be left exposed to the public, strong consideration should be given to assigning a security guard to watch over the reel. An occasional investment of $75 or $100 to avoid theft of an expensive reel of cable, to say nothing of the disturbance to the job if the cable is gone when installers arrive to begin work in the morning, is a wise investment

Exact terminating locations must be identified in advance. This obvious though often overlooked step can lead to costly mistakes if forgotten. The cable must be pulled all the way to where it will be terminated, not just into the room. The designer should provide a fold-flat diagram (which results from folding out the walls of the room as if they were the sides of a box hinged at the intersections of the walls and the floor) for each terminating room as a part of the initial planning walkout.

ADEQUATE DUCT SPACE

Current industry practice, outside telephony, tends not to utilize available duct space very efficiently. As a result, one often finds a 1/2-inch fiber cable as the only cable in a 4-inch duct. As a matter of good design, cable should fill approximately 60 to 70 percent of the available duct space.

Since it is difficult to pull cable into a duct that is already occupied, it is very important to detail, on the print, where the different media will be pulled. Since most fiber cables are 3/4-inch or smaller in diameter, for efficient use, the larger 3- or 4-inch diameter conduits should be subdivided with innerducts.

Innerduct is flexible, nonmetallic conduit that is pulled in multiples into the main conduit. It is made of polyethylene and comes with a pull rope already installed within it. Innerduct serves a threefold purpose:

- Subdividing the main duct
- Protecting the fiber cable
- Reducing friction

The inside of the innerduct is designed to let the cable glide smoothly within it. This is accomplished by the use of special coatings and by the physical properties of the inner wall. Corrugated innerduct is the most popular form because of its flexibility. A 3-inch conduit can be subdivided with up to four 1-inch innerducts; up to six 1-inch innerducts can be placed in a 4-inch conduit.

Innerduct not only allows the initial fiber installation to be effectively accomplished, but allows expansion capabilities to be built into the system. This can provide great return on investment if there is any expectation of future expansion of the fiber system.

INITIAL PLANNING WALKOUT

Prior to the start of cable pulling, a planning walkout should be performed by the project engineer, the lead technician, and the project superintendent. The project engineer, responsible for system design, will be well aware of the customer's needs. The lead technician, an experienced fiber installer who will not necessarily be the one who will do the installation, has the responsibility to plan and to answer as many questions as possible before the crew is on site. The superintendent can provide general project information and needs to know as much as possible about the installer's activities to coordinate with other trades and activities in the vicinity of the installation.

The walkout should start at one termination site (Figure 12-1). This location will be designated with a clearly stated name, such as "#12 Communication Room." A rough fold-flat sketch will identify the fiber terminal location and locations of other major system components, as well as other relevant features in the room. The floor and walls will be marked with a pencil to indicate approximate locations of terminations. Notes should be made on the sketch concerning conduit type and size, as well as whether pull string or measurement/pull tape has been used. Verification of whether multiple ducts are involved is requested from the superintendent.

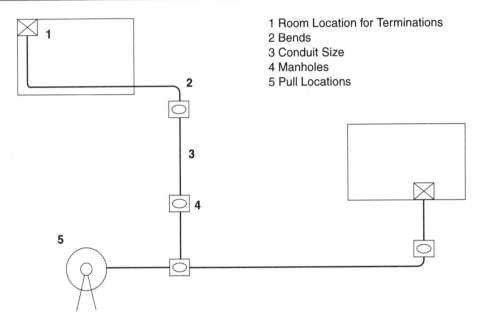

1 Room Location for Terminations
2 Bends
3 Conduit Size
4 Manholes
5 Pull Locations

Figure 12-1 Preplanning walkout sketch.

Identification should be made defining which conduits go where either by visual inspection or by tugging on the pull tape. The path of the conduit should then be walked, making note on the diagram of any pull boxes, manholes, or other abnormalities. This process must be carried out for each run on the job.

The more information the engineer provides on the print, the faster the installation will progress. Remember, the object of this planning walkout is to prevent the entire pulling crew from being idle while the lead technician tries to track down the engineer or superintendent for clarifications or additional information.

REVIEW QUESTIONS

1. The most accurate method of measuring for conduit pulls is _____
   a. from a set of prints.
   b. field measurements taken with a wheel.
   c. measuring with a pull tape.

2. How much extra cable must be added to make a fusion splice in a man-hole?
   a. 30 feet
   b. 1 percent
   c. 30 feet for each side of the splice
   d. 10 feet

3. Subdividing larger conduits with innerducts _____
   a. efficiently uses duct space.
   b. allows for future expansion.
   c. offers additional protection to the fiber optic cable.
   d. all of the above

4. The purpose of a planning walkout is to _____
   a. get a rough idea of the installation.
   b. measure the conduit run.
   c. clarify all the final details.
   d. none of the above

13

FIBER OPTIC
CABLE PLANT
DOCUMENTATION

JIM HAYES

Documenting the fiber optic cable plant is a necessary part of the design and installation process for the fiber optic network. Documenting the installation properly as part of the planning process can save time and material in the installation and allow better planning for upgrading. During installation, it will speed the cable pulling and installation since the routing and terminations are already documented. After installation, the documentation can be completed with test data for acceptance by the end user. If equipment is repositioned on a network, as is always the case, proper documentation will allow easier rerouting to the proper end points. And during troubleshooting, proper documentation is mandatory for tracing links and finding faults.

The documentation process for fiber optics, however, is different from that of most copper cable installations. Where a copper cable of Cat 3 or 5 type is generally used to connect a single link, fiber optic cables—especially backbone cables—may contain many fibers connecting a number of different links that may not even be going to the same place. Furthermore, copper cable may be tested only for continuity while loss data will be required for most fiber optic cable plants.

The fiber optic cable plant, therefore, must be documented as to the path of every fiber, connection, and test. Data that should be kept include the following:

- Cable: manufacturer, type, length
- Splice and termination points (at distance markers)
- Fiber: fiber type and size, splice and connection data, losses
- Connections: types (splice or connectors and types), fibers connected, losses
- Paths: where the link path goes in every cable

Most of this data can be kept in a database that stores component, connection, and test data. Long links may also have OTDR data that can be stored as printouts or in special file formats for later viewing in case of problems. If the OTDR data is stored digitally, a database of data files should be kept to allow finding specific OTDR traces more easily.

CABLE PLANT RECORD KEEPING

Good cable plant record keeping can be used for many purposes.

Designing the Cable Plant

By starting with a simple graphic package showing the spatial relationship of the interconnections, you can design the layout of the entire cable plant; assign cable, fiber, and patch panel designations; and provide reports for each of the components and quantity used. For every cable you should specify the type of cable, installation method, number of fibers, types of fibers, and the length of the run.

Bidding the Installation Job

Good records and reports will give you all the information you need for estimating the total material content of the job and the length of cable runs. Individual cable reports will give you the details on special requirements for installation (direct burial, aerial, conduit, and so on) to help price labor costs and the total job.

Installing the Cable Plant

Records should provide the information you need to send materials to the proper location, determine what needs to be installed where, and make the proper interconnections. Patch panel printouts can be left taped inside the panel covers to provide local documentation of proper connections for checking during troubleshooting.

Testing the Cable Plant

After the cable plant is installed, keep test data on each individual fiber run. You can test any link or end-to-end loss and store it as a permanent record associated with that fiber link. You can compare the data if you retest at a later date or have problems requiring troubleshooting or restoration. Note how the test was performed and what instruments were used. Use a power meter and source to get accurate end to end loss data and only use an OTDR if you need to verify splice loss (requiring bidirectional OTDR measurements) or to insure no bending losses were induced during installation.

Troubleshooting Cable Plant Problems

A well-documented cable plant will be much easier to troubleshoot. The documentation will tell you where cables go, what they are connected to, and how far they run between points, making problem areas easier to find. Test data will indicate any degradation over time that needs to be addressed.

Documenting the Cable Plant for Customer Acceptance

Most customers require documentation of the cable plant before acceptance. Good record keeping from the beginning, including all the data on the cable plant and final testing, will require only a copy as a report to the customer for acceptance. If customers wish to maintain their own cable plants, they can use the same data for future reference.

THE DOCUMENTATION PROCESS

Before you start entering data, you should have a basic layout for the network completed. A sketch may work for a small building network but a large campus or metropolitan network will probably need a complex computer-aided design (CAD) layout. The best way to set up the preliminary data is to use a facility drawing showing the locations of all cables and connection points. Identify all the cables and closets/panels and then you are ready to transfer this data to a database.

You must know where all the cables go in the network and what every fiber will be connected to. You should know the specifications on every cable and fiber: what types of fiber are being used, how many fibers, cable construction type, estimated length, and installation technique (buried, aerial, plenum, riser, etc.)

It will help to know what types of panels and closet hardware are being used, and what end equipment (if any) is to be connected. If you are installing a big campus cable plant with many dark (unused) fibers, some will probably be left open at the panels, and that must be documented also. When designing a network, it is a very good idea to have spare fibers and interconnection points in panels for future expansion, rerouting for repair, or moving network equipment (Figure 13-1).

05/19/2000   Cable List Detailed Report   page 1

   Cable ID: C001   Length: 1130 m   Fibers: 8

   Type: MM 62.5   Mfg: NLC   Color: Orange

   Notes: Pulled through 4 in PVC duct in steam line tunnels to Library.
   8 fibers, 900 um buffer, color coded.

Cost Estimates:

Cable _____

Installation _____

Termination _____
   (if included with cable)

Panel A	* Fiber ID *	Notes	Panel B
ADM01-01-0001	C001-0001	red	SCA01-01
ADM01-01-0002	C001-0002	orange	SCA01-02
ADM01-01-0003	C001-0003	yellow	SCA01-03
ADM01-01-0004	C001-0004	blue	SCA01-04
ADM01-01-0005	C001-0005	green	SCA01-05
ADM01-01-0006	C001-0006	violet	SCA01-06
ADM01-01-0007	C001-0007	black	SCA01-07
ADM01-01-0008	C001-0008	white	SCA01-08

Figure 13-1 Sample of a cable report format.

PROTECTING RECORDS

Cable plant documentation records are very important. Keep several backup copies of each document, whether it is stored in a computer or on paper, in different locations. In fact, keep several copies on disks and on paper in different physical locations in case of disasters. If a copy is presented to the customer, the installer should maintain his or her own records for future projects. Access to modify records may be restricted using passwords if one wishes to stop unauthorized changes to the documentation.

REVIEW QUESTIONS

1. What data should be included with the documentation?

   a. _____

   b. _____

   c. _____

   d. _____

   e. _____

2. The first step in documenting a cable plant is _____
   a. a database.
   b. a sketch.
   c. test data.
   d. OTDR traces.

14

ESTIMATING AND BIDDING FIBER OPTIC INSTALLATION

F. DOUGLAS ELLIOTT AND
PAUL ROSENBERG

THE ART OF ESTIMATING

In the beginning of this chapter we are concerned with the art of estimating, which may include estimating a building's communications system, LANs, or any other similar project. We look at the tools, the skills, the experience, and the background that the estimator needs to do the job. In this chapter we generalize the procedures and use generic materials if required.

Anyone involved with or considering the art of estimating must realize from the outset that there are certain skills that require proficiency before the job should be attempted. If you do not have these skills and are not willing to develop them, you could be heading for big problems. You may get away with a bluff on the first several jobs, but eventually ineptitude will give you away.

SKILLS

Knowledge of Mathematics

You must have a good knowledge of mathematics. Let me qualify this statement by saying that many a job has been lost by improperly adding a column of figures. Many jobs have been taken because a decimal point has been misplaced somewhere along the line or figures have been transposed incorrectly. A quantity may not have been properly calculated or extended into the estimate itself. When the final figures are incorrect, you are in a costly mess.

Command of the English Language

You must have a good command of the English language and a phenomenal knowledge of acronyms and abbreviations. Remember, you are probably not the only person who will read your estimates. You might be called away on business and someone else may have to complete your project. You will, in all probability, have to hand over your calculations and computations to the project manager in order to control the hours, check your labor units, material lists, and so on. If your colleagues cannot understand your writing and/or mathematical figures on the pages, how can they ever hope to work with them? There are hundreds of acronyms and abbreviations associated with communications and fiber optics. You must be fluent in them.

Imagination

You must have a vivid and skilled imagination. You have to be able to imagine or visualize what the finished project or specific portions of that project will actually look like. Imagination is an acquired skill and is developed with experience. If you are mounting a distribution/patch panel on a wall in a telecom closet, you must be able to visualize the finished item in order to do that job properly. Too many people lack this ability and miss out on major components or labor units. It is hard to express the amount of imagination that may be required for you to do a proper and complete job, but please believe me, it is required. Those of us who lack this ability should try drawing small detail sketches in order to ensure that they have included all pertinent materials. This can also be an advantage if you have the same type of repetitive installation on several floors of a building.

Understanding of On-Site Expenses

If you do not know the project and the industry very well, you cannot hope to estimate it properly. If you are not aware of the procedures for installing connectors, making splices, installing the cables (both copper and fiber), testing requirements and procedures, and all the other details of the installation, you cannot estimate the job. You have to be aware of the different product lines out there, and how they are to be installed and protected. If you are not familiar with the time needed and the care that must be exercised in the installation of these products, you cannot estimate the job. I find it very difficult to understand how a non-tradesperson or a novice estimator in communications and fiber optics can possibly hope to properly and completely quote a job and show a profit. If this happens, then there must be a lot of blind luck and prayer involved.

Handwriting Skills

Another area that must be considered when you are estimating is your writing skills. If your handwriting is like a doctor's, print everything, as most estimators do. The reason is simple: People must be able to read what is on the paper. This is especially true in three or four months, or a year or two later, when you have to go back over a project, buy materials, make final plans for the installation, and schedule deliveries and workers.

TOOLS

I make sure that the materials and tools that I use when estimating are of the very highest quality for reliability and consistency. Some of these items you may not think of as tools, but they are the tools of the estimator's trade.

I use a pencil for the simple reason that everyone makes mistakes; it is much easier to erase pencil than ink.

Your adding machine or calculator should only be the type that can provide you with printout or tape. Pocket calculators do not work. They have a very limited means of storing information, if any at all, and thus no means for you to review or crosscheck your calculations.

Tape recorders come in very handy when going out on a site visit, or for that matter when you think of something that would be of use later. Either writing the thought down or recording it on tape saves it for future consideration. The site visit can be a disaster without a tape recorder. When you talk into a tape recorder on the site, the tour will all come back to you when you play back your notes. Hopefully, you will not miss out on the important aspects of the project. Certainly you are still going to have to take measurements and so on, but a lot of the detail can be saved very accurately on a tape recorder.

Another tool that comes in very handy is a camera. If you are working on an existing building and you want to have a permanent record of some site details, a camera will do the trick. On a new installation within an existing building, or within a building that is being constructed from the ground up, a picture can be worth a thousand words, and may show details that the human eye can miss. Be sure that you ask permission to use a camera on any site—owners get very nervous and upset if they think you are an industrial spy.

Estimating sheets can be as simple as a few lines on a piece of paper with a heading, or as fancy as you want to get, with company logos and letterhead. Estimating sheets are available from many stationery suppliers in a loose leaf pad form (Figure 14-1). From left to right: the first column lists the item number; column two lists the quantity of the items; column three lists the name and short description of item; column four lists the individual item cost; column five lists cost of the individual items (col. 3) multiplied by the quantity of the item (col. 2). Column six lists the labor cost per unit item, and column seven lists the labor cost per unit (col. 6) multiplied by the quantity of the item (col. 2). The bottom line of the page is for the dollar totals in columns 5 and 7.

Figure 14-2 shows a summary sheet or a recap sheet. It is exactly the same type of calculation sheet, although the headings are different. Each individual page of the estimate and the individual estimating group divisions (e.g., group 1 cable, group 2 connectors, etc.) should be listed here. In the case of a smaller project, the individual pages can sometimes be listed so that you can get a good overview of the whole project. Column 7 can be used for factoring purposes—if the cable installation listed on page one will be done from a ladder more than 5 feet and less than 10 feet high, you may want to increase the labor cost 5 percent to cover this problem. Most estimating manuals will list approximate factoring percentages.

Some people would rather use the computerized estimating programs on the market today, and I find them very satisfactory. However, they do not replace the estimator. The estimator still must work up the labor units and the proper description, extension, and so on, of the job. The use of a computerized program, although it may ease the pain slightly, is not the total answer. There is still a tremendous amount of leg work that must be done in order to do a proper and complete estimating job.

SITE VISIT

Just as skill in your particular job or being very competent with your trade is important, so the site visit or project tour is every bit as important. The site visit will give you insight into what may confront you when you eventually come onto the job site. You may see obstructions to getting your equipment onto the site. The whole job may have to be done from 12-foot ladders instead of rolling scaffold. You will not know unless you look.

On a site visit it is necessary to take in and digest everything that you possibly can. You are on the site to equip yourself with some background so that you can properly and completely put together an accurate package for the estimate. A site visit will show your actual equipment locations, routing for your conduits, and proper elevations. It will allow you to apply the architectural drawings that usually accompany the bid package. If the drawing package is incomplete, complete it. Also, during site visits, be especially aware of construction in progress or

EXAMPLE 1 ELECTRIC LTD.
ANYWHERE U.S.A. 45678
UPS & DOWNS TOWERS PROJECT
CABLE REQUIRED

			Initials	Date
	Prepared by		PDS	4/95
	Approved by		DAP	5/95

#	QTY.	DESCRIPTION	COST PER UNIT / UNIT	TOTAL COST	LABOR PER UNIT	LABOR TOTAL COST	#
1							1
2	100'	6c DISTRIBUTION	24⁸⁰	2480⁰⁰	12/M	600⁰⁰	2
3		CABLE					3
4							4
5							5
6							6
7							7
8							8
9							9
10							10
11							11
12							12
13							13
14							14
15							15
16							16
17							17
18							18
19							19
20							20
21							21
22							22
23							23
24							24
25							25
26							26
27							27
28							28
29							29
30							30
31							31
32							32
33							33
34							34
35							35
36							36
37							37
38							38
39							39
40							40
41							41
42							42
43							43
#		PAGE TOTALS		2480⁰⁰		600⁰⁰	

Figure 14-1 Sample estimation sheet.

EXAMPLE 1 ELECTRIC LTD.
RECAP SHEET
UPS & DOWNS TOWERS PROJECT

	Initials	Date
Prepared by	PDS	4/95
Approved by	DAP	5/95

	PAGE	DESCRIPTION	MATERIAL COST	LABOR COST	TOTAL COST	FACTOR PERCENT	
1							1
2							2
3							3
4							4
5							5
6	1	CABLE REQ'D	2480⁰⁰	600⁰⁰	3080⁰⁰	5%	6
7							7
8							8
9							9
10							10
11							11
12							12
13							13
14							14
15							15
16							16
17							17
18							18
19							19
20							20
21							21
22							22
23							23
24							24
25							25
26							26
27							27
28							28
29							29
30							30
31							31
32							32
33							33
34							34
35							35
36							36
37							37
38							38
39							39
40							40
41							41
42							42
43							43
#		PAGE TOTALS	2480⁰⁰	600⁰⁰	3080⁰⁰		

Figure 14-2 Sample summary or recap sheet.

planned changes that may affect your estimating. Changes in construction or location of buildings or equipment, as well as changes in elevations, can cause big changes in estimates.

FIBER OPTIC INSTALLATIONS—THIS IS REALITY

When we deal with a fiber optic communications project, we will be looking in most cases at an existing office building or factory that is to be retrofitted and/or upgraded in order to provide the customer with a state-of-the-art, high-speed communications system that will satisfy needs now and well into the future. What we have to do is to change the often unstructured wiring system into a structured wiring system that will comply with federal, state, and local codes. We may also have to deal with the Telecommunications Industry Association/ Electronics Industries Alliance (TIA/EIA) standards. We must calculate what we are going to need in the way of hardware and the locations for it.

In most instances, the communications room, wiring or telecom closet (whatever you care to call it), in existing buildings is totally inadequate. It was probably inadequate when the building was being built 25 or 30 years earlier, but even more so now. Many of the communications rooms were shared rooms. Not only was the telephone equipment mounted there, but so was the lighting and power transformer and electrical distribution panel. No wonder there is trouble with interference in the communications systems!

Most users have a number of different types of data and voice cabling already installed: thicknet and thinnet Ethernet coax, twinax for IBM systems, bus and tag for mainframes, unshielded twisted pair (UTP), shielded twisted pair (STP), and voice-grade wire. Much of that will not work with the high-speed equipment that they are now purchasing. Thus, they are starting to look at and go with fiber. They are looking at retrofitting in a big way. They want a system that will be free from radio frequency interference (RFI) and electromagnetic interference (EMI) interference. The only way to do this is with fiber optics. With fiber optics, they have a good clean clear signal that they are making good use of. Therefore, the project will probably be to remove all the existing copper cabling that is used for communications and replace it with fiber optics.

A site visit to a plant such as that can be absolutely frightening. You would have to be looking at routing. Can you use the existing conduits? What are the temperature conditions? Are you are going to have high heat problems? What are the dirt conditions? What about winter cold temperatures? In other words, what are all of the contingencies that we may be dealing with? All of a sudden we have a nightmare, or do we? That nightmare can make you a lot of money if you are careful, good, observant, and have the skills to present a good proposal.

Good planning and proper reuse of the existing facilities can save the customer money and make your bid much more competitive. Remember that you can pull a lot more fiber optic cable (and have a lot more capacity) in a conduit than you can UTP or coax. Thus proper distribution of the backbone/riser cable will greatly reduce cost. Now take this idea, put it to paper, and work it up. Make the customer an offer that he or she cannot refuse.

Treat estimating as a skilled art. Do not cut corners. Utilize the most up-to-date materials, tools, and computer programs. Use every one of your skills at hand to your advantage. Be informed and keep informed. Technology is changing as you read this book. Set aside time to read and study your craft. Take pride in your work and enjoy it. Do not simply endure your work or you will never be a success.

ESTIMATING PROCESS FOR OPTICAL FIBER INSTALLATIONS

The first step in estimating is to ascertain the overall requirements of the job. You must get a clear picture in your mind of how this job will flow; and more important, where will the money come from, and when. In addition, you must understand the scope of the work you are quoting on, and exactly what will be required of you. These are primary concerns, and are the first considerations in any good estimate. In order to verify that all of these factors are considered, many estimators use checklists that they review for every project. You should develop your own.

The three-part estimating process that I recommend is fairly standard:

1. The takeoff
2. Writing up the estimate
3. Summarizing the estimate

The process of "taking off" the job is literally taking the information off the plans and transferring it to separate sheets. The estimator interprets the graphic symbols on the plans and translates them into words and numbers, which can then be processed.

In writing up the estimate, the estimator transfers the takeoff information to special sheets, assigns both material and labor prices to each item, and totals these prices on each sheet.

In the summary, the estimator adds all the pricing sheets to give a total cost for the material and labor for the project. Any other costs that will be required for the project's completion, overhead for the maintenance of the company's internal operations, and last, profit must also be included to determine a final bid price.

Every piece of material that will be used on the installation in question must be itemized; each item must be assigned a material cost and a labor cost. You then multiply these costs by the quantities and add them all up to arrive at a base cost. From base cost, you add overhead, profit, and other job expenses to arrive at a selling price. The differences between applying this process to optical fiber cables and copper conductor cables are as follows.

Fragility delays. Since optical fibers are vulnerable to damage, they must be handled differently than copper cables. You must use higher labor units for these cables than you might use for similar-sized copper cables. The estimating process can be handled in much the same way as other cables, but the labor units that you use will need to be a bit higher (10–20%), to account for the extra care. Because of this fragility (note that the cables are very tough, often tougher than copper cables. The fibers, however, are very fragile, being made of glass). The cables must attach to the pulling line differently (the pulling tension can be applied to only one part of the cable, a layer called the "strength member"), and cannot be pulled with too much tension. Pulling too hard on an optical fiber cable will tend to either pull the layers apart or damage the optical fibers.

Testing. Optical cables require quite a bit more testing than copper cables, with the exception of Cat 5 UTP used in high-speed networks. The general testing procedures that the cables require (including inspecting the cable upon delivery) are included in the basic labor rates given in Table 14-1. In addition to this, you should include extra testing time for each fiber installed. Note that we are talking each and every fiber, not each cable. Since one cable may contain hundreds of fibers, there is a very big difference in terms; make sure that you account for every length of fiber.

The Takeoff

The process of taking off high-tech systems is essentially the same as the process used for conventional estimating. By taking off, we mean the process of taking information off a set of plans and/or specifications, and transferring it to estimate sheets. This requires the interpretation of graphic symbols on the plans, and transferring them into words and numbers that can be processed. Briefly, the rules that apply to the takeoff process follow.

1. Review the symbol list. This is especially important for high-tech work. High-tech systems are not standardized and therefore vary widely. Make sure you know what the symbols you are looking at represent. This is fundamental.
2. Review the specifications. Obviously it is necessary to read a project's specifications, but it is also important to review the specifications before you begin your takeoff. Doing this may alert you to small details on the plans that you might otherwise overlook.
3. Clearly and distinctly mark all items that have been counted. Again, this is obvious, but a lot of people do this rather poorly. This must be done in such a way that you can instantly ascertain what has been counted. This means that you should color every counted item completely. Do not just

put a check mark next to something you counted; color it in so fully that there will never be any room for question.

4. Always take off the most expensive items first. By taking off the most expensive items first, you are assuring that you will have numerous additional looks through the plans before you are done with them. Very often, you will find stray items that you missed on your first run through. You want as many chances as possible to find all of the costly items. This way, if you make a mistake, it will be less expensive.

5. Obtain quantities from other quantities whenever possible. For example, when you take off conduit, you do not try to count every strap that will be needed. Instead, you simply calculate how many feet of pipe will be required and then include one strap for every 7–10 feet of pipe. We call this obtaining a quantity from a quantity. Do it whenever you can. It will save you good deal of time.

6. Do not rush. Cost estimating, by its very nature, is a slow, difficult process. In order to do a good estimate, you must do a careful, efficient takeoff. Do not waste any time, but definitely do not go so fast that you miss things.

7. Maintain a good atmosphere. When performing estimates, it is very important to remain free of interruptions and to work in a good environment. Spending hours counting funny symbols on large, crowded sheets of paper is not particularly easy; make it as easy on yourself as you can.

8. Develop mental pictures of the project. As you take off a project, picture yourself in the rooms, looking at the items you are taking off. Picture the item you are taking off in its place, its surroundings, the things around it, and how it connects to other items. If you get in the habit of doing this, you will greatly increase your skill.

Labor Units

The labor units shown in Table 14-1 are necessarily average figures. They are based on the following conditions:

1. An average worker
2. A maximum working height of 12 feet
3. A normal availability of workers
4. A reasonably accessible work area
5. Proper tools and equipment
6. A building not exceeding three stories
7. Normal weather conditions

Table 14-1 Fiber Optics Labor Units

Labor Item	Labor Units (Hours)	
	Normal	Difficult
Optical fiber cables, per foot:		
1–4 fibers, in conduit	0.016	0.02
1–4 fibers, accessible locations	0.014	0.018
12–24 fibers, in conduit	0.02	0.025
12–24 fibers, accessible locations	0.018	0.023
48 fibers, in conduit	0.03	0.038
48 fibers, accessible locations	0.025	0.031
72 fibers, in conduit	0.04	0.05
72 fibers, accessible locations	0.032	0.04
144 fibers, in conduit	0.05	0.065
144 fibers, accessible locations	0.04	0.05
Hybrid cables:		
1–4 fibers, in conduit	0.02	0.025
1–4 fibers, accessible locations	0.017	0.021
12–24 fibers, in conduit	0.024	0.03
12–24 fibers, accessible locations	0.022	0.028
Testing, per fiber	0.12	0.24
Splices, including prep and failures, trained workers:		
fusion	0.30	0.45
mechanical	0.40	0.50
array splice, 12 fibers	1.00	1.30
Coupler (connector-connector)	0.15	0.25
Terminations, including prep and failures, trained workers:		
polishing required	0.40	0.60
no-polish connectors	0.30	0.45
FDDI dual connector, including terminations	0.80	1.00
Miscellaneous:		
cross-connect box, 144 fibers, not including splices	3.00	4.00
splice cabinet	2.00	2.50
splice case	1.80	2.25
breakout kit, 6 fibers	1.00	1.40
tie-wraps	0.01	0.02
wire markers	0.01	0.01

Any set of labor units must be tempered to the project to which it is applied. It is a starting point, not the final word. Difficult situations typically require an increase of 20 to 30 percent. Some very difficult installations may require even more. Especially good working conditions or especially good workers may allow discounts to the labor units of 10 to 20 percent, possibly more in some circumstances.

Charges

Training

Many high-tech installations require you to teach the owners or their representatives to use the system. This is a difficulty peculiar to high-tech work, since you do not have to spend time teaching building owners how to use conventional electrical items, but you will certainly have to spend time teaching them how to use a sound system. Not only that, you may have to supply operating instructions, teach a number of different people, and answer numerous questions over the phone after the project is long over.

The question here is whether you include training charges in special labor or include these costs as a separate job expense. This decision is essentially at the discretion of the estimator. It is, however, usually best to charge general training to job expenses, and to include incidental training in special labor.

If you come from a background in the electrical construction industry, make sure that you accept these expenses as an integral part of your projects. Do not avoid them. The people who are buying your systems need this training, and they have a right to expect it. Include these costs in your estimates, and choose your most patient workers to do the training.

Overhead

What percentage of overhead to assign to any type of electrical work (and how to assign it) can be a hotly debated subject. Everyone seems to have his or her own opinion. Whatever percentage of overhead you charge, consider raising it a bit for high-tech projects. As we have already said, the purchasing process is far more difficult for high-tech work than it is for other more established types of work. In addition, there are a number of other factors that are more difficult for high-tech work than for more traditional types of work.

Almost every factor we can identify argues for including more overhead charges for high-tech work. Not necessarily a lot more, but certainly something more. When we contract to do high-tech installations, we are agreeing to go through uncharted, or at least partially uncharted, waters. This involves greater risk. And if we do encounter additional risk, it is only sensible to make sure we cover these risks. We do this by charging a little more for overhead and/or profit.

REVIEW QUESTIONS

1. List the five skills that are important to estimating.
    1. _____
    2. _____
    3. _____
    4. _____
    5. _____

2. The three standard parts of estimating are
    1. _____.
    2. _____.
    3. _____.

3. What is the tallest building assumed by the labor units?
    a. two stories
    b. three stories
    c. four stories
    d. five stories

4. What is the normal labor unit for testing 12 fibers?
    a. .12
    b. 1.44
    c. .44
    d. .24

5. Labor units in difficult situations typically increase _____
    a. 5–10 percent.
    b. 10–20 percent.
    c. 20–30 percent.
    d. 40 percent or more.

15

FIBER OPTIC CABLE PULLING

THOMAS A. DOOLEY AND
JERALD R. ROUNDS

(with hints from Northern Lights Cable)

Electrical wire installers know how to pull cable. The basic approach to pulling fiber optic cable differs little from the techniques used to pull copper or aluminum. However, just as aluminum responds differently than copper when pulled, fiber has its own idiosyncrasies. The focus of this chapter is as much on what to avoid as on how to pull fiber optic cable (Figure 15-1).

AVOIDING DISASTER

The first step in pulling cable is to measure and cut the material. Inaccurate measurements are a disaster in fiber cable installation. Splices are much more critical with fiber than with metal cable because a minimum loss budget must be maintained and splices cause loss. Thus, assumptions and guess work are simply not allowed.

The physical characteristics of fiber cable must always be borne in mind during the installation process. The two characteristics that are particularly important during the installation process are *tensile stress* (pulling load) and *bending radius*.

Figure 15-1   Fiber optic cable is often installed in conduit for outside plant applications.

The glass fiber within the cable is fragile, and although the cable has been designed to protect the fiber, it is more easily damaged than metal cables and requires greater care during the process of cable pulling. You simply cannot afford to break fiber cable during the pulling process.

Damage to cable can come in many forms. The most common form of damage, a broken fiber, is also the most difficult to detect. In addition to fracture, fiber can be cracked from too much tension. As a result, no gorillas should be allowed on the cable installation crew.

DESPOOLING CABLE

Although the optical cordage may outwardly resemble copper cordage, the two are significantly different. A failure of optical cordage may occur when improper methods of pulling and despooling are employed. Pulling the outer jacket will cause a compression of the optical fiber and cause significant attenuation increase. This condition once initiated is usually irreversible. One should also avoid cable twist when despooling fiber optic cable to prevent stressing the fibers.

Longitudinal force on the jacket can cause temporary elongation and subsequent fiber compression. Therefore, cable should be reeled off the spool, not spun over the edge of the spool. This will eliminate cable twist, which will make coiling much easier (Figure 15-2).

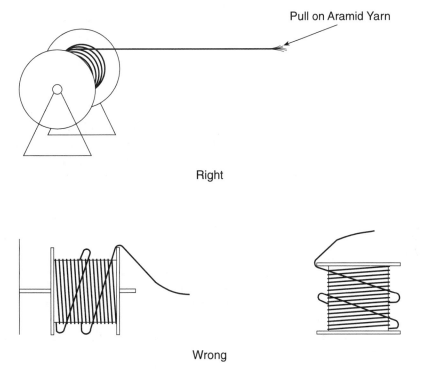

Figure 15-2   Despooling fiber optic cable.

When unreeling the cordage, tension should be applied only to the strength member. The strength member and buffered fiber do not stretch. If the outer jacket is used to unreel the cordage from the spool, the resulting shock tension on the outer jacket will allow the jacket to stretch momentarily. The jacket will then return to its normal state. Therefore, the fiber and strength member may be compressed in the retraction of the outer jacket. This will cause macrobend attenuation in the cable.

PULLING FORCE

The pulling force must be kept below a designated limit for the specific cable being installed. This is usually 600 pounds for outside plant (OSP) cable and 300 pounds or less for other cables. The pulling force must also be kept uniform.

Most fiber cable cannot handle high impact loads, so the cable should not be jerked. Included within the cable is a strength member, which is purposely placed there to facilitate installation. This member, not the glass fiber, must always be used when tension is to be placed on the cable.

When using power equipment to pull OSP cable, tension monitoring equipment or breakaway swivels must always be used. Power equipment must never be used on inside fiber because the allowable pulling force is so small.

Testing

In order to avoid quality problems after installation, as well as to eliminate disputes that could arise over responsibility for damaged cable, testing of cable prior to installation is recommended. Preinstallation testing becomes particularly important under certain circumstances, such as installation under especially difficult conditions, expensive cable being installed, or an unknown supplier or manufacturer of the cable.

Preinstallation testing need not be complex or time-consuming. If the cable shows no signs of damage, it can be tested with a continuity tracer. If all fibers transmit light, it is highly likely to be good cable. If there is even a hint of possible damage to the cable, it must be tested or outright rejected.

Postinstallation testing of cable (preceding termination) is recommended if any abnormal circumstances were encountered during the installation process. Examples of such abnormal circumstances might include exceeding the allowable pulling tension during the pull or sheath damage observed during or after the pull. Any time there is a possibility of damaged cable, the sooner it is detected and remedied the better.

BENDING FIBER TOO TIGHTLY

The second most common problem is bending the fiber on too tight a radius. The bending radius is always important in a static condition. However, it becomes even more important under tensile loading, because the tensile stresses due to bending are added to those due to pulling. A minimum bending radius of 10 cable diameters must be maintained over long-term, static conditions.

When cable is placed under a tensile load while being pulled, a minimum of 20 cable diameters is recommended. It should be noted that a design in which a cable is placed by hand into a tray allows a tighter radius than one where installation will be carried out by pulling the cable in.

INTERFERENCE WITH OTHER INSTALLATIONS

Another source of damage to cable during the installation process is interference with other installations. Careful coordination must be carried out in order to give maximum protection to the cable. This might mean that fiber cable should go in first, with other cable placed carefully over the top to afford some protection to the fiber cable in the event other contractors might later access the same tray.

Probably more common is the situation where the fiber cable installer wants to be the last person to place material in the duct, so that other installations of more robust materials placed by craftspersons not sensitive to the fragile nature of fiber cable do not damage the fiber cable.

Yet another potential source of damage to fiber cable is that caused by sharp corners or protrusions, such as where conduit enters pull boxes and cabinets. These are commonly found in the working environment of a construction project and must be avoided or negated by the use of innerduct. If multiple layers of installations are possible or potentially hazardous obstacles are in the path of the cable, installing the fiber cable in corrugated innerduct may be a good idea. The innerduct protects the fiber cable and its distinctive orange color helps others notice it.

In circumstances where these types of damage are liable to occur, such as with rough buss duct or conduit that must be field cut and fabricated, a little investment in inspection prior to installation could save significantly in terms of both time and money if obstacles can be detected and eliminated.

PROCEDURES FOR PULLING CABLE

As with any cable-pulling operation, set up the reel so cable pays off the top (see Figure 15-2). Place the reel as close as possible to the conduit or innerduct opening. Lubricant is recommended on all but the shortest of pulls. Be sure to use lubricant appropriate for fiber optic cable.

For long runs, the operation must be accomplished in two or more stages (as shown in Figures 15-3 and 15-4). A pull box or manhole is placed as close as possible to the middle of the run, or at reasonable pulling intervals along the run for longer runs.

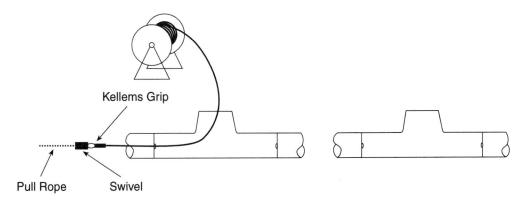

Kellems Grip

Pull Rope   Swivel

Figure 15-3   The first half of a multistage pull.

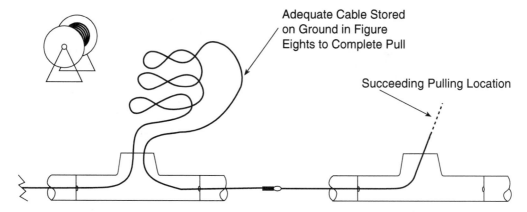

Figure 15-4 The second half of a multistage pull.

The pull is started at the middle pulling location and proceeds in both directions. A pulling eye should be attached to the cable's strength member or a Kellems grip (similar to the Chinese finger puzzle) should be placed on the end of the cable and connected to the pull rope through a swivel. The first part of the pull is then carefully made, pulling adequate spare cable beyond the end of the run.

If more than two runs are required, enough cable is pulled each time to enable reaching the full length of run on that side of the pulling location and the spare cable is stored in figure eights on the ground.

Once the end of the run is reached in one direction, the process begins again at the center of the run, pulling in the opposite direction. Sufficient cable is paid off the reel and laid on the ground in figure eights to reach the other end of the run. The cable is then pulled through successive pulling locations, storing the excess cable in figure eights at each location. Remember to place the figure eights in a safe area, well away from traffic.

Cable with aramid yarn as a strength member can be attached to a pulling eye directly as shown in Figure 15-5. If attaching the Kellems grip to the cable, first remove the last 2 feet of sheath, fiber, and antibend rod, leaving only the Kevlar pulling yarn. Then, slide the grip onto the next 2 feet of sheathed cable. Attach the pulling swivel to the Kellems grip loop and tie the leading 2 feet of Kevlar to the pulling swivel. In this way, the pulling load is distributed between the sheath and the Kevlar strength member. On cable using fiberglass, Kevlar, or stainless steel embedded within the sheath, simply put the grip on the sheath. A cutback is not necessary since this special sheath acts as the strength member. The last step in the attachment process is to wrap the installed grip with vinyl tape, starting on the cable and working up to, but not including, the swivel.

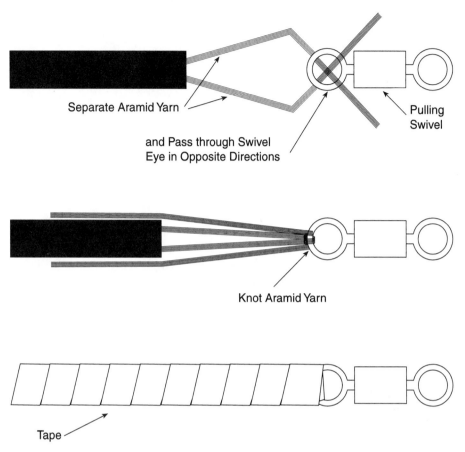

Separate Aramid Yarn

and Pass through Swivel
Eye in Opposite Directions

Pulling
Swivel

Knot Aramid Yarn

Tape

Figure 15-5   Attaching pulling swivel to cable strength member.

Simplex and duplex cable should always be pulled using a pulling rope and swivel whenever possible. Should it become necessary to continue to pull on the jacket, a mandrel should be used (Figure 15-6). Using a 6-inch diameter mandrel, wrap five turns. Tape or allow a finger to maintain tension on the first wrap from the loose end. This will insure that force is transferred to the aramid yarn strength member in the same way a ship's capstan allows rope to be pulled with no attachment to the capstan except friction.

Before beginning the pull, make sure you have not tied the cable in a knot or looped any other cable. Start the cable into the innerduct or conduit slowly at first to make sure that everything is going as planned. After the amount of cable

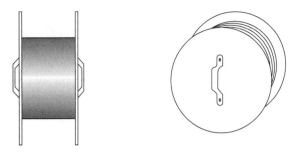

Northern Lights Cable, Inc. Pulling Spool

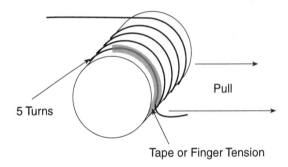

Pull

5 Turns

Tape or Finger Tension

Figure 15-6   Using a mandrel for pulling cable.

that will be handled by the pullers at the other end of the run has entered the duct, apply the lubricant. Stop the pull, make a quick funnel out of paper and pour about 50 percent of the lubricant needed on the pull into the feed end of the duct. Resume pulling, increasing pulling speed. Add the remainder of the lubricant as needed. If the pulling crew has to handle lubricated cable, but does not want to take the time to clean the cable off, latex gloves work great.

When pulling with rope, maximum speed through the duct should be about 3 feet per second, or 2 miles per hour. When mule tape is used, the speed can be tripled. This is because at speeds higher than 3 feet per second, rope will cut grooves in conduit bends, but mule tape will not.

Rack the cable after the entire pull is complete (Figure 15-7). Protect the cable within a manhole or pull box with innerduct if the manhole or pull box is congested or will be in the future. Also protect the cable if you will be pulling

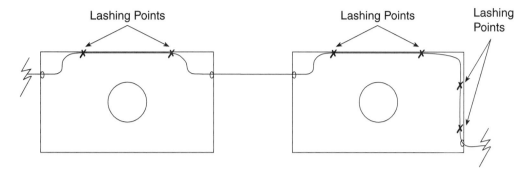

Figure 15-7 Racking cable in manholes.

through multiple manholes or pull boxes. Start racking at the center manhole or pull box and work toward the ends. Use double-looped cable ties, cinched up tight on OSP cable, but not tight enough to indent the sheath on indoor type cable.

Place cable identification tags on cable or innerduct at every location that humans could possibly visit in the future. This is important because of the relative ease with which a fiber cable carrying thousands of customers can be cut by someone who does not recognize fiber optic cable.

Cable tags should be plastic, about 2 inches by 3 inches, marked with indelible marker and should state the following information:

- Fiber size, such as 62.5, 62.5/125
- Where fiber is accessible on both ends, such as Term Room 410 to Term Room 912
- Who owns or is responsible for the fiber, such as Telecom System Department

If lubricant dripping out of the conduit will be a safety or aesthetic problem, seal the cable within the duct with a mechanical squeeze plug, such as the one manufactured by Jack Moon Industries, or use a suitable canned spray foam. The former is neat, quick, clean, removable, and expensive; the latter is not!

HOLDING CABLE FOR STRIPPING

When stripping the jacket and buffer materials from cord ends of very short pieces, it is advisable to thread the cordage through the fingers to grip the inner buffer and optical fiber. This method allows minimum force to be used during buffer stripping without allowing the buffer in short cords to be pulled out (Figure 15-8).

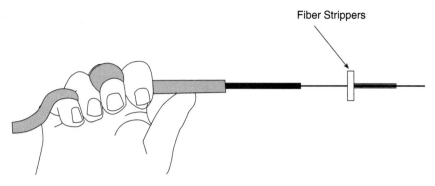

Figure 15-8 Holding cable for stripping fiber.

REVIEW QUESTIONS

1. Fiber is pulled on the _____
   a. buffer.
   b. strength member.
   c. fiber.
   d. binding tape.

2. The pulling force for outside plant cable is usually _____
   a. 300 pounds.
   b. 400 pounds.
   c. 500 pounds.
   d. 600 pounds.

3. The minimum bending radius for cable being pulled is _____
   a. 30 times the cable diameter.
   b. 20 times the cable diameter.
   c. 10 times the cable diameter.
   d. 5 times the cable diameter.

4. For long runs _____
   a. the cable should be spliced.
   b. the pull should be split up into two or more stages.
   c. greater force can be used.
   d. a mandrel wrap should be used.

16

FIBER OPTIC RESTORATION

LARRY JOHNSON

Fiber optic technology has provided communications users with the ability to transmit longer distances with increased amounts of information. This has paralleled the growth in communications provided by computers, fax machines, cellular services, and video using digital transmission systems. Fiber optics has been used mostly in backbones due to its capability to handle vast amounts of information, whether from city to city or file servers to users in campus applications.

As the growth of these industries increases, so does our reliance on the fiber systems being used for the transmission. Telephone companies and other service providers have routines and disciplines well established for restorations. Their long spans in exposed aerial and direct buried applications make restoration practices critical to the operation of their communications systems.

Local area networks (LANs) are unique in that they consist of intra- and interbuilding links of relatively short distances when compared to the various types of wide area and metropolitan networks using optical communications. This requires different approaches and equipment to be responsive to emergency restorations.

PROACTIVE PLANNING VERSUS REACTIVE RESTORATIONS

All networks start at the conceptual design stage. During this stage we must establish a value to the type of posture we should take for providing for the physical plant and its protection. Since the establishment of the Fiber Distributed Data Interface (FDDI) standard in the late 1980s, we have seen a serious effort to provide protection in both the network and physical routing of optical communications systems. Why? FDDI was designed as a backbone system operating at 100 Mb/s. With this large amount of data, users wanted reliability to assure that their entire data systems would not fail in case of either node or cable failure. FDDI's counterrotating ring offered route redundancy, optical bypass switches for network protection, and diagnostics for system management.

Today's designers should learn from these lessons. Networks with high data rates, critical reliability, security, and priority users should be designed for route diversity. Route diversity means two specific different routes (not putting two cables in the same duct). This, of course, can create a cost issue in both materials and construction. For example, if a campus is built on a system of steam tunnels in a star topology, the expense would be high to create a physical ring. But the tradeoffs between network reliability and cost must be made at the design stage.

Types of Faults

Another issue to review in the design stage is the types of failures that have occurred in the past. History repeats itself, even in network failures. Tables 16-1 and 16-2 show some of these failures.

A quick review of these tables allows us to identify certain areas and types of faults.

A: Patch Panel Related

These failures occur around the patch panel. The cause could be improper dressing of the jumpers and cables, improper keying of connectors, contamination of the connection, or improper cable routing and localized damage.

B: System Related

Over- or underdriving the optical transmission still causes either total or intermittent failure.

C: Installation Related

Improper bend radius, clamping of the cables, and improper rolls of the transmit/receive fibers are common types of problems. Installations occurring around previously installed fiber networks can also create failures due to lack of attention in dressing, termination, and cable routing.

Table 16-1 Typical Cable System Faults

Fault	Cause	Equipment	Remedy
Bad connector	Dirt or damage	Microscope	Cleaning/ polishing/ retermination
Bad pigtail	Pigtail kinked	Visual fault locator	Straighten kink
Localized cable attenuation	Kinked cable	OTDR	Straighten kink
Distributed increase in cable attenuation	Defective cable or installation specifications exceeded	OTDR	Reduce stress/ replace
Lossy splice	Increase in splice	OTDR	Open and redress
	Loss due to fiber stress in closure	Visual fault locator	
Fiber break	Cable damage	OTDR Visual fault locator	Repair/replace

Table 16-2 Typical Causes of Failures in LANs

No.	Failure Location	Type of Failure
1	Broken fibers at connector joints	A
2	Broken fibers at patch panels	A
3	Cables damaged at patch panels	A
4	Fibers broken at patch panels	A
5	Cables cut in ceilings and walls	D
6	Cables cut through outside construction	D
7	Contaminated connections	A
8	Broken jumpers	A
9	Too much loss	B
10	Too little loss (overdriving the receiver)	B
11	Improper cable rolls	C
12	Miskeyed connectors	C
13	Transmission equipment failure	B
14	Power failure	B

D: Construction Related (Major and Minor)

Normally due to construction and work-related activities, this could be caused by backhoes and other heavy construction equipment digging up cables. Cables can be damaged in aerial plant due to improper installation techniques, falling branches, automobile crashes, gunshots, and lightning. In LANs, cuts through walls and ceilings, mistaken cutting of cables, improper clamping, and breaking fibers at the connectors by improper handling or by accident are examples of localized failures due to poor cable identification or lack of care by workers.

EQUIPMENT USED IN RESTORATION

1. **Fiber optic cleaning kit.** Many faults with optical systems are caused by contamination of connectors, so simple cleaning will resolve these problems. Remember to keep connectors clean and capped when not in use. (Resolves problem 7 [Table 16-2])

2. **Optical inspection scope.** Used to identify poor connector finishes and surface contamination or damage. Magnification can be from 30 to 400 power. The microscope needs to have an interface to hold the connector type securely for viewing. (Identifies problem 7 [Table 16-2])

3. **Optical loss test set or test kit.** Used for end-to-end loss tests. The set consists of a stabilized light source and a calibrated power meter. The instruments should match the operating wavelength, fiber type, and connector of the transmission system. The power meter by itself is the essential go–no go instrument in fiber optic troubleshooting. This instrument allows us to check power levels at the transmitter, receiver, and at any connection point in a system to isolate whether the problem is with the electro-optics (low transmitter power or proper receiver power indicating electronic problems) or the fiber optic cable plant (increased loss beyond allowable margins). Using these instruments and good system documentation, the user should be able to diagnose almost every network fault. (Identifies power levels with problems 1–6, 8–11, and 13 [Table 16-2])

4. **Visual tracers and fault locators.** Inexpensive instruments that transmit visible light through a fiber. The more powerful versions use red He, Ne, or diode lasers operating in the visible spectrum and can locate breaks though many types of jumpers and buffered fibers, as long as the buffer or cable jacket is translucent. These instruments will not work on most multifiber cables or black- or gray-jacketed single-fiber cables. White light and red LED versions are also available but do not have the power to locate internal breaks. (Identifies problems 1–3, 8, 11 [Table 16-2])

5. **Optical Time Domain Reflectometer (OTDR).** The instrument everyone often considers first in restorations is the most application dependent. In long-distance networks where most outages occur far away from the end equipment, the OTDR is critical for a restoration posture. In the case of LANs, most problems cannot easily be identified with the OTDR, since its distance resolution makes most short cables impossible to view. In a LAN, cable plants must be diagnosed through the use of the above mentioned types of equipment, unless one is dealing with a campus network where cables are longer than 500 to 1,000 meters. Due to the high cost of an OTDR, rental may be better than purchasing one.

IDENTIFYING THE PROBLEM

The first requirement for restoration is, of course, to identify the problem. To guide you through that process, a flowchart is shown in Figure 16-1. First one must determine if the problem is in the electronics or the fiber optic cable plant. Using the power meter, unplug the connector at the receiver, measure the power at the receiver, and compare it to the network specification. To measure the receiver power, it is sometimes necessary to run equipment diagnostics at the transmitter to output a test signal. If the power level is within specification, there is most likely a problem with the transmission electronics. If this is the case, you must call in the network specialists to diagnose the network problem.

If the receiver power is low, go to the transmitter and measure the power output there, using a known good jumper cable and the power meter. If the power is low, there is a transmitter problem that must be fixed. If the transmitter power is within specification, there is too much loss in the cable plant. This must be diagnosed.

Diagnosing the cable plant can start with a visual continuity check with a visual tracer or fault locator. If the cable plant has just been installed or rerouted at patch panels, be certain that the correct connections have been made at every patch panel. This requires starting at one end and tracing the fiber path through every connection. Even if you do not find a misconnection, you will find where the signal stops, thus identifying the segment of the cable plant with the problem.

Several problems can cause the faulty segment, including a cable cut or stress causing loss, broken or damaged connectors, or a bad splice. If only one fiber is affected and there are no splices in the cable, the problem is most likely a damaged connector. Examining it visually for kinking of the cable or other damage, or examining the end of the connector ferrule with the microscope may show the fault. If the cable is short, like most LAN cables, a visual tracer or fault locator can be used most effectively. The most effective tool is good cable plant records, however, since cable cuts or damage caused by construction or erroneous cutting can be found by following the cable route and looking for obvious damage.

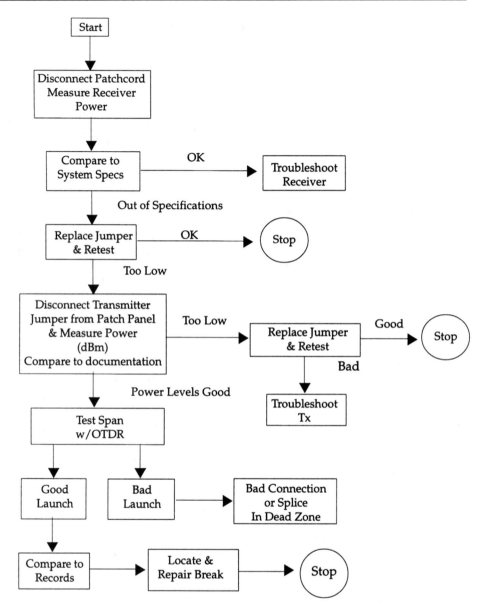

Figure 16-1 Problem identification flowchart. Courtesy The Light Brigade

If the cable is long (over 500 meters or 1,640 feet) or underground, an OTDR may be necessary to find and diagnose the problem. But since the OTDR has poor short-length resolution, it will not be useful in finding many LAN problems such as broken connectors on jumper cables or failures in short links. If you are using an OTDR to find faults in multimode links, use the 850-nm range since it has much higher resolution than the 1300-nm range and is thus more likely to find problems.

RESTORING SERVICE

Once the fault and location have been identified, we need to restore the outage. This sounds simple but can be complex. However, if the problem is with only one fiber in the cable, switch to a spare fiber and replace the bad connector as time permits. If immediate emergency restoration is necessary, consider these issues.

Basic Questions

1. Does the span have retrievable slack? If so, we can pull the slack back and make one repair point.
2. Will it be easier, quicker, and/or less expensive to replace the section versus repairing the section?
3. Will we install connectors or use splices for the repair?
4. Either way we will need to protect the repair point(s) to prevent future damage. This could mean adding closure(s), patch panel(s) or rerouting of cable to other physical points.
5. Can the system handle the additional losses caused by the additional connectors, splices, and fiber length?
6. If necessary, can we provide a drop cable over the ceiling, in the ceiling, down the roof, and so on until we can make a permanent restoration?

WHAT ABOUT THOSE STORAGE LOOPS?

Networks and the buildings housing them must accommodate many adds, moves, and changes over their life spans. Designers planning for these also resolve one of the many headaches associated with LAN cable restorations. This is the issue of retrievable slack versus nonretrievable slack.

Retrievable Slack

Cable spans designed with slack points allow for spare cable to be pulled together, allowing for only one termination point. The use of quick mechanical splices or direct connectorization allows for restoration to take place with minimal losses. Because most cables within the building are of the breakout or distribution styles, these allow for easy retermination. The main issue is how to store the splice/connection panel or miniclosure. These products provide strain relief of the cable and physical protection of the splice or connectors.

Placement could be above the floor, wall mounted, or ceiling mounted. In most of these situations aesthetics and size will be key factors. For many users, security and access may need to be considered. The cable should be strain relieved and prepared, leaving as much slack as possible for future entry and possible rerouting for future changes. (Hopefully we can use this fiber and reroute for future needs as well.)

EMERGENCY RESTORATION (WITH RETRIEVABLE SLACK)

The cable fault has been found and the spare has been pulled back to the failure point. We must confirm that the cable break is where it appears to be. The use of visual light sources should be used to check each fiber from both ends. (We would not want to have a second break point 1 foot away and not cut it out.)

The site must be checked to find the best point and method to repair the fibers. The cables may be pulled back to a ceiling, floor, post, or other location for physical mounting. This location should be noted on your drawings and documented. The panel should also be labeled and possibly secured. After the cables are repaired, the fiber spans should be retested for loss using the optical loss test set.

Nonretrievable Slack

Let us take a serious look at our options. Will it be quicker to pull in a new cable or segment? Should I install a new segment? Should I splice or connectorize and how do I protect these? Without retrievable slack we must add a section of cable to the span. This will require not only two termination points but also twice the labor and material. We must also have a length of fiber equal to or greater than the amount of fibers in the span. As you can see, the price for not leaving slack is both more expensive and labor consuming. The process for the determinations is the same as the retrievable slack, except now we have two points that need to be addressed.

FIBER OPTIC RESTORATION FOR SINGLEMODE NETWORKS

Networks involving singlemode fibers tend to be more critical and more difficult to repair than their multimode counterparts. This is due to the higher speeds and longer distances encountered. In addition, singlemode systems tend to be directly buried, placed by aerial methods, and in longer conduit systems. These techniques usually provide an optimum amount of protection for the optical cable. However, when they fail it is normally catastrophic. Natural disasters, construction activities, and accidents cause most of these failures.

They cause greater outages affecting more users, more lost revenue, and are a magnitude more complex to restore. This is why singlemode users tend to have a restoration plan in place and use slack times to practice the plan.

Summary of Suggestions

1. Prioritize your fibers. Most networks have spare fibers. When your system is down, get the priority fibers fixed first, then worry about the spares.
2. Have spare connectors and a connectorization kit or mechanical splice kit available.
3. Have a trained crew that knows how to terminate and test the fiber network. Have emergency phone numbers available for access.
4. Have the proper test equipment and tooling for the job.

Planning for Restoration

The Basic Recommended Restoration Posture

All users of communications systems must have a basic posture to address what would happen should a failure occur. Following are several recommendations.

1. All fiber routes should be properly documented for both optical performance and physical routing.
2. This should include patch panel designations, signal type, and interconnect routing information.
3. All transmitters and receivers should be documented to their optical transmit and power levels. Receivers should be documented for both minimum and maximum power levels.
4. All spans should be documented for optical loss. This would normally be at both 850/1300 nm for multimode applications and 1310/1550 nm for singlemode applications. The documentation should identify the fiber size and manufacturer.
5. If OTDR tests have been performed, copies of the OTDR traces should be included in the test reports.
6. If cable has sequential markings, the difference between the markings tells us the actual cable length in meters or feet for each segment. This should be documented.
7. Fibers should be identified and prioritized to allow for priority fibers to be quickly restored.

Restoration Planning

Imagine a system failure and having to restore a damaged optical cable. Let us look at some of the issues that would need to be addressed.

1. How would we know of the problem?
2. Who is first advised of the outage?
3. Is there a technical team on call to respond?
4. Is this a dedicated route without backup or alternative routing? If yes, this would require emergency restoration.
5. If not, this could be a planned restoration. Planned restorations allow for more flexibility, providing better quality in restoration.
6. Do you have records including OTDR prints, optical power levels, up-to-date "as built" drawings on all segments?
7. Do you have an emergency restoration program?
8. Do you have emergency restoration kits?
9. Have these been evaluated with your staff including management, engineering, construction, and maintenance?
10. Are your customers prioritized? Are there any contracts and/or services that could affect priorities (e.g., emergency services, government, military)?
11. Do you have prioritized fibers?
12. What is the time allowance for restoration?
13. Is this a restoration in which we will allow compromises on splice loss to bring the system up and will resplice later when better prepared?
14. What is the maximum allowable splice loss for restorations?
15. Is there a vehicle available that can allow a team to work within it so that they can work in a well-lit, dry environment with a power supply?
16. Are all necessary materials and equipment easily accessible by the team en route to the outage?
17. How many splice and test sets do you have?
18. If using a fusion splicer, do you know the fiber types and settings for the equipment?
19. What is the OTDR with the highest resolution?
    What pulse width?
    Wavelength?
    Do you know the index of refraction of the fibers and manufacturers in your system?
20. What type of communications will be used between OTDR operators and splicers?
21. What is the limitation of this equipment?
22. What level of heavy equipment will be necessary?
23. How do you determine the location of a cable cut? Can you triangulate?

24. In the case of a single cut with retrievable slack, what equipment will be used?

25. In the case of a cut without retrievable slack, what equipment will be used?

    Team A _____

    Team B _____

    Which team is quicker? _____

    Which team has the most experience? _____

26. In the case of massive cable failure, how many cables can you repair simultaneously?

27. Can this restoration be done safely or will we be delayed?

28. Is there anything we can do about this?

29. Where is spare cable stored and how is it identified?

30. What else can go wrong?

31. Have we missed anything? Equipment? Environment? Staff? Tools?

Restoration (Miscellaneous Issues)

1. How do we keep the restoration plan and staff current?

2. Have you graded your staff on fiber optic restoration abilities?

3. Do you have annual/semiannual procedures for testing/evaluating existing dark fibers?

4. How are test reports filed?

5. Where are test reports filed?

6. What about updates?

7. Do these include locations where cable slack is stored?

8. Each cable segment should be evaluated for worst case failures. Has this been done?

9. Do the emergency restoration kits (ERK)s include a bill of materials/checklist of all tools and components and suppliers?

10. Do you have adequate amount of inventory and consumables? Are any of these date coded?

11. Do you photograph/film your restorations? The use of film and/or pictures provides a good learning and review tool. In the case of litigation the pictures can be invaluable.

Postrestoration Recommendations

1. Redocument and retest your splices, spans, and segments.

2. Adjust your "as built" drawings. New vaults, closures, splices, and slack cable points may need to be added or adjusted.

3. Schedule and conduct a meeting to review all aspects of the restoration.
   A) What happened? What were the cause and impacts?
   B) What did we do well?
   C) What did not work? (Technique, equipment, products, staff)
   D) How can this be resolved?
   E) How can we improve?
   F) What needs to be done to rebuild kits and replenish inventory?

REVIEW QUESTIONS

1. A high loss splice is found with a(n) _____
   a. OTDR.
   b. microscope.
   c. visual fault locator.
   d. cable identifier.

2. Faults can be generalized into four categories:
   1. _____
   2. _____
   3. _____
   4. _____

3. Poor connectors are found with a(n) _____, and are best remedied by _____
   a. visual fault locator, retermination.
   b. visual fault locator, cleaning.
   c. microscope, cleaning.
   d. microscope, retermination.

4. The best was to find a break in an accessible LAN cable is with a(n) _____
   a. OTDR.
   b. plant records.
   c. visual fault locator.
   d. power meter.

5. If two fibers in a 12-fiber cable show high loss, while the other 10 show low loss, the most likely cause is _____
   a. improper bend radius.
   b. improper pulling method.
   c. kinked cable.
   d. dirty connectors.

17

FIBER OPTIC TESTING

JIM HAYES

OVERVIEW

Testing fiber optic components and systems requires making several basic measurements. The most common measurement parameters are shown in Table 17-1. Optical power—required for measuring source power, receiver power, and loss or attenuation—is the most important parameter and is required for almost every fiber optic test. Backscatter and wavelength measurements are the next most important, and bandwidth or dispersion are of lesser importance. Measurement or inspection of geometrical parameters of fiber are essential for fiber manufacturers. And troubleshooting installed cables and networks is required.

Standard Test Procedures

Most test procedures for fiber optic component specifications have been standardized by national and international standards bodies, including the EIA in the United States and International Electrotechnical Commission (IEC) or International Organization for Standardization (ISO) internationally. Procedures for measuring absolute optical power, cable and connector loss, and the effects of many environmental factors (such as temperature, pressure, flexing, etc.) are covered in these procedures.

Table 17-1 Fiber Optic Testing Requirements

Test Parameter	Instrument
Optical power (source output, receiver signal level)	Fiber optic power meter
Attenuation or loss of fibers, cables, and connectors	Fiber optic power meter and source, test kit or optical loss test set (OLTS)
Source wavelength*	Fiber optic spectrum analyzer
Backscatter (loss, length, fault location)	OTDR
Fault location	OTDR, visual cable fault locator
Bandwidth/dispersion* (modal and chromatic)	Bandwidth tester or simulation software

* Rarely tested in the field

In order to perform these tests, the basic fiber optic instruments are the fiber optic power meter, test source, OTDR, optical spectrum analyzer, and an inspection microscope. These and some other specialized instruments are described below.

Fiber Optic Instrumentation

Fiber Optic Power Meters

Fiber optic power meters (Figure 17-1) measure the average optical power emanating from an optical fiber and are used for measuring power levels and, when used with a compatible source, for loss testing. They typically consist of a solid state detector (silicon [Si] for short wavelength systems, germanium [Ge] or indium-gallium arsenide [InGaAs] for long wavelength systems), signal conditioning circuitry and a digital display of power. To interface to the large variety of fiber optic connectors in use, some form of removable connector adapter is usually provided.

Power meters are calibrated to read in linear units (milliwatts, microwatts, and nanowatts) and/or dB referenced to one milliwatt or one microwatt optical power. Some meters offer a relative dB scale also, useful for laboratory loss measurements. (Field measurements more often use adjustable sources set to a standard value to reduce confusion. See section entitled Fiber Optic Test Sources).

Power meters cover a very broad dynamic range, over 1 million to 1. Although most fiber optic power and loss measurements are made in the range of 0 dBm to −50 dBm, some power meters offer much wider dynamic ranges. For testing analog CATV systems or fiber amplifiers, special meters are needed with extended high power ranges up to +20 dBm (100 mW).

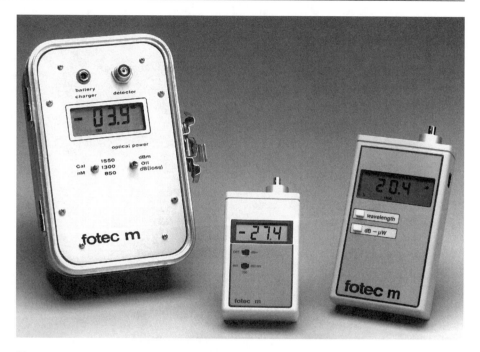

Figure 17-1 Fiber optic power meters come in rugged, mini, and handheld packages. Courtesy Fotec, Inc.

Fiber optic power meters have a typical measurement uncertainty of ±5%, when calibrated to transfer standards provided by national standards laboratories such as the U.S. National Institute of Standards and Technology (NIST). Sources of errors are the variability of coupling efficiency of the detector and connector adapter, reflections off the shiny polished surfaces of connectors, unknown source wavelengths (since the detectors are wavelength sensitive), nonlinearities in the electronic signal conditioning circuitry of the fiber optic power meter, and detector noise at very low signal levels. Power meters with very small detectors may have two problems that cause measurement errors. The light from the fiber may overfill the detector or the detector may saturate at high power levels. Since most of these factors affect all power meters, regardless of their sophistication, expensive laboratory meters are hardly more accurate than the most inexpensive handheld portable units.

Fiber Optic Test Sources

In order to make measurements of optical loss or attenuation in fibers, cables, and connectors, one must have a standard signal source as well as a fiber optic power meter. The source (Figure 17-2) must be chosen for compatibility with the

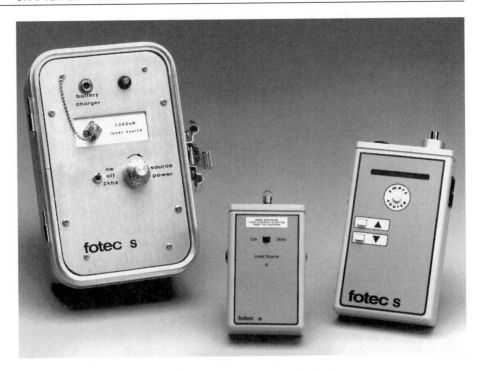

Figure 17-2 Fiber optic test sources. Courtesy Fotec, Inc.

type of fiber in use (singlemode or multimode, with the proper core diameter) and the wavelength desired for performing the test. Most sources are either LEDs or lasers of the types commonly used as transmitters in actual fiber optic systems, making them representative of actual applications and enhancing the usefulness of the testing.

Typical wavelengths of sources are 665 nm (plastic fiber), 850 nm (short wavelength multimode glass fiber), and 1300 and 1550 nm (long wavelength multimode and singlemode fiber). LEDs are typically used for testing multimode fiber and lasers are used for singlemode fiber, except for the testing of short singlemode jumper cables with LEDs. The broad spectral output of an LED has higher attenuation in singlemode fiber than a laser, causing significant errors on cables longer than about 5 kilometers. The source wavelength can be a critical issue in making accurate loss measurements on longer cable runs, since attenuation of the fiber is wavelength sensitive, especially at short wavelengths. Thus all test sources should be calibrated for wavelength.

Adaptability to a variety of fiber optic connectors is important also, since over 70 styles of connectors exist, although the types most commonly used are ST and SC for multimode fiber and SC or FC for singlemode fiber. The new small form factor connectors (LC, MT–RJ, VF–45) are also gaining in popularity. Some LED sources use modular adapters such as power meters to allow adaptation to various connector types. Lasers almost always have fixed connectors. If the connector on the source is fixed, hybrid test jumpers with connectors compatible with the source on one end and the connector being tested on the other must be used.

Optical Loss Test Sets/Test Kits

The optical loss test set (Figure 17-3) is an instrument formed by the combination of a fiber optic power meter and source that is used to measure the loss of fiber, connectors, and connectorized cables. Early versions of this instrument were called attenuation meters. A test kit has a similar purpose, but is usually comprised of separate instruments and includes accessories to customize it for a specific application, such as testing a fiber optic LAN, telco, or CATV.

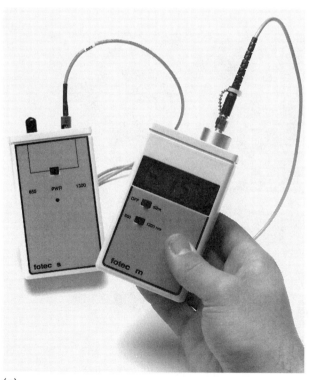

(a)

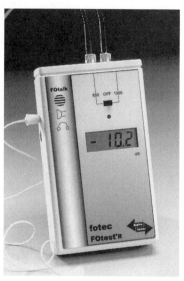

(b)

Figure 17-3 An optical loss test set can be separate source and power meter (a) or an integrated single instrument (b). Courtesy Fotec, Inc

The combination optical loss test set (OLTS) instrument may be useful for making measurements in a laboratory, but in the field, individual sources and power meters are more often used, since the ends of the fiber and cable are usually separated by long distances, which would require two OLTSs, at double the cost of one fiber optic power meter and source. And even in a laboratory environment, several different source types may be needed, making the flexibility of a separate source and meter a better choice.

Optical Time Domain Reflectometers

The OTDR (Figure 17-4) uses the phenomenon of fiber backscattering to characterize fibers, find faults, and optimize splices. Since scattering is one of the primary loss factors in fiber (the other being absorption), the OTDR can send out into the fiber a high-powered pulse and measure the light scattered back toward the instrument. The pulse is attenuated on the outbound leg and the backscattered light is attenuated on the return leg, so the returned signal is a function of twice the fiber loss and the backscatter coefficient of the fiber.

If one assumes the backscatter coefficient is constant, the OTDR can be used to measure loss as well as to locate fiber breaks, splices, and connectors. In addi-

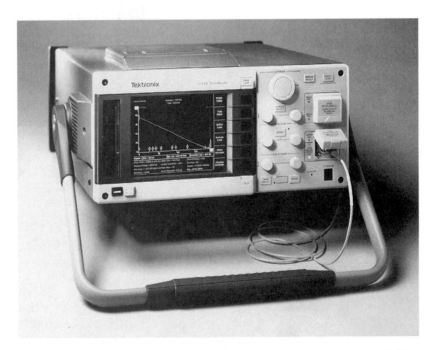

Figure 17-4a Full featured OTDRs offer maximum range and flexibility.
Courtesy Tektronix

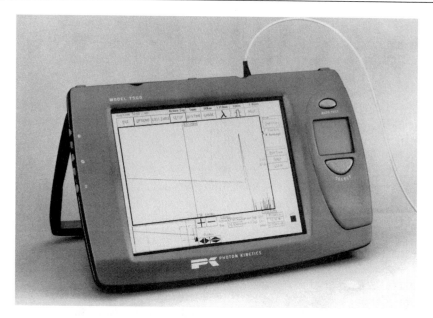

Figure 17-4b   Mini OTDRs offer fewer features in much smaller packages and at less cost. Courtesy Photon-Kinetics, Inc.

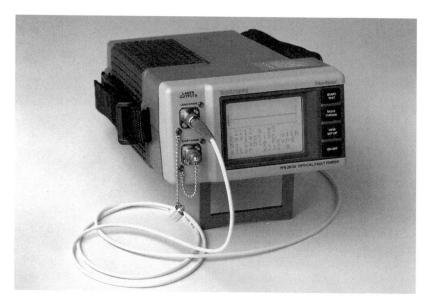

Figure 17-4c   Fault finders are single OTDRs for troubleshooting. Courtesy Tektronix

tion, the OTDR gives a graphic display of the status of the fiber being tested. And it offers another major advantage over the source/fiber optic power meter or OLTS in that it requires access to only one end of the fiber.

OTDRs generally are used to test all outside plant installations, especially to confirm the loss of splices between lengths of cables. The distance resolution of a typical OTDR, however, is too long to see the typical patch cords used in most cable plants, limiting their usefulness for premises cabling installations. Also, all network specifications call for testing the cables with a source and power meter, as that tests the cables exactly as they will be used in the application.

The uncertainty of the OTDR measurement is heavily dependent on the backscatter coefficient, which is a function of intrinsic fiber scattering characteristics, core diameter, and numerical aperture. It is the variation in backscatter coefficient that causes many splices to show a "gain" instead of the actual loss. Tests have shown that OTDR splice loss measurements may have an uncertainty of up to 0.8 dB. OTDRs must also be matched to the fibers being tested in both wavelength and fiber core diameter to provide accurate measurements. Thus many OTDRs have modular sources to allow substituting a proper source for the application.

While most OTDR applications involve finding faults in installed cables or optimizing splices, they are very useful in inspecting fibers for manufacturing faults. Development work on improving the short-range resolution of OTDRs for LAN applications and new applications such as evaluating connector return loss promise to enhance the usefulness of the instrument in the future.

OTDRs come in three basic versions. Full-size OTDRs offer the highest performance and have a full complement of features such as data storage, but are very big and high priced. Mini OTDRs provide the same types of measurements as a full OTDR, but with fewer features to trim the size and cost. Fault finders use the OTDR technique, but are greatly simplified to provide just the distance to a fault, to make the instruments more affordable and easier to use.

Visual Cable Tracers and Fault Locators

Many of the problems in fiber optic networks are related to making proper connections. Since the light used in systems is invisible, one cannot see the system transmitter light. By injecting the light from a visible source, such as an LED or incandescent light bulb, one can visually trace the fiber from transmitter to receiver to insure correct orientation and also to check continuity. The simple instruments that inject visible light are called visual fault locators (Figure 17-5).

If a powerful enough visible light, such as an HeNe or visible diode laser, is injected into the fiber, high loss points can be made visible. Most applications center around short cables such as those used in telco central offices to connect to the fiber optic trunk cables. However, since visible light covers the range where OTDRs are not useful, it is complementary to the OTDR in cable troubleshoot-

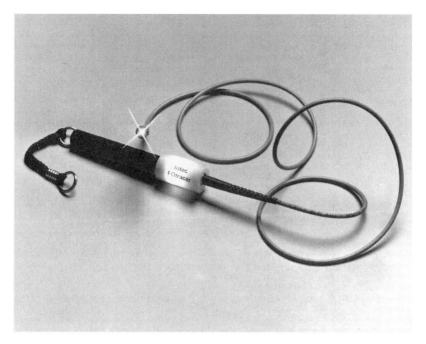

Figure 17-5   A simple fiber tracer is a flashlight coupled to the fiber optic cable. Courtesy Fotec, Inc.

ing. This method will work on buffered fiber and even jacketed single-fiber cable if the jacket is not opaque to the visible light. The yellow jacket of singlemode fiber and orange of multimode fiber will usually pass the visible light. Most other colors, especially black and gray, will not work with this technique, nor will most multifiber cables. However, many cable breaks, macrobending losses caused by kinks in the fiber, bad splices, and so on can be detected visually. Since the loss in the fiber is quite high at visible wavelengths, on the order of 9 to 15 dB/km, this instrument has a short range, typically 3 to 5 kilometers.

Fiber Identifiers

If one carefully bends a singlemode fiber enough to cause loss, the light that couples out can also be detected by a large area detector. A fiber identifier uses this technique to detect a signal in the fiber at normal transmission wavelengths. These instruments usually function as receivers, able to discriminate between no signal, a high speed signal, and a 2 kHz tone. By specifically looking for a 2 kHz "tone" from a test source coupled into the fiber, the instrument can identify a specific fiber in a large multifiber cable, especially useful to speed up the splicing or restoration process.

Fiber identifiers can be used with both buffered fiber and jacketed single-fiber cable. With buffered fiber, one must be very careful to not damage the fiber, as any excess stress here could result in stress cracks in the fiber that could cause a failure in the fiber anytime in the future.

Optical Continuous Wave Reflectometers

The optical continuous wave reflectometer (OCWR) was originally proposed as a special purpose instrument to measure the optical return loss of singlemode connectors installed on patch cords or jumpers. Unfortunately, its purpose became muddled between conception and inception. As actual instruments came on the market, they had much higher measurement resolution than appropriate for the measurement uncertainty (0.01 dB resolution versus 1 dB uncertainty), leading to much confusion on the part of users as to why measurements were not reproducible. In addition, several instruments were touted as a way to measure the optical return loss of an installed cable plant, obviously in ignorance of the fact that they would also be integrating the backscatter of the fiber with any reflections from connectors or splices. Since the measurement of return loss from a connector can be made equally well with any power meter, laser source, and calibrated coupler, and an OTDR is the only way to test installed cable plants for return loss, the OCWR has little use in fiber optic testing.

Visual Inspection with Microscopes

Cleaved fiber ends prepared for splicing and polished connector ferrules require visual inspection to find possible defects. This is accomplished using a microscope with a stage modified to hold the fiber or connector in the field of view (Figure 17-6). Fiber optic inspection microscopes vary in magnification from 30 to 800 power, with 30 to 100 power being the most widely used range. Cleaved fibers are usually viewed from the side, to see breakover and lip. Connectors are viewed end-on or at a small angle to find polishing defects such as scratches (Figure 17-7).

Fiber Optic Talksets

Although technically not a measuring instrument, fiber optic talksets are sometimes used for fiber optic installation and testing. They transmit voice over fiber optic cables already installed, allowing technicians splicing or testing the fiber to communicate effectively. Talksets are especially useful when walkie-talkies and cellular telephones are not available, such as in remote locations where splicing is being done, or in buildings where radio waves will not penetrate.

Attenuators

Attenuators are used to simulate the loss of long fiber runs for testing link margin in network simulation in the laboratory or self-testing links in a loopback config-

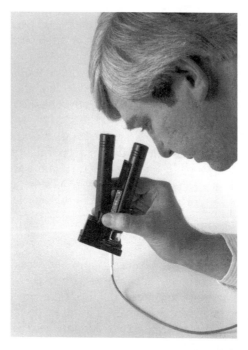

Figure 17-6 Microscopes allow inspection of connectors for polish quality, cleanliness, and faults. Courtesy Fotec, Inc.

FIBER OPTIC FINISH EVALUATION

 TYPICAL 3
MICRON FINISH AT 200X

 CRACKED AND
CHIPPED FINISH AT 200X

 TYPICAL 1
MICRON FINISH AT 200X

 PLUCKED
FINISH AT 200X

 TYPICAL 0.3
MICRON FINISH AT 200X

 CRACKED AND
PLUCKED FINISH AT 200X

Figure 17-7 Connector faults are easily seen through a microscope. Courtesy Buehler LTD

uration. In margin testing, variable attenuators are used to increase loss until the system has a high bit error rate. For loopback testing, an attenuator is used between a single piece of the equipment's transmitter and receiver to test for operation under maximum specified fiber loss. If systems work in loopback testing, they should work with a proper cable plant. Thus, many manufacturers of network equipment specify a loopback test as a diagnostic/troubleshooting procedure.

Attenuators can be made by gap loss, or a physical separation of the ends of the fibers, inducing bending losses or inserting calibrated optical filters. Both variable and fixed attenuators are available, but variable attenuators are usually used for testing. Fixed attenuators may be inserted in the system cables where distances in the fiber optic link are too short and excess power at the receiver causes transmission problems.

Test Jumper Cables and Bulkhead Splice Adapters

In order to test cables using the FOTP-171 insertion loss test, one needs to establish test conditions. This requires launch jumper cables to connect the test source to the cable under test and receive cables to connect the fiber optic power meter. For accurate measurements, the launch and receive cables must be made with fiber and connectors matching the cables to be tested. To provide reliable measurements, launch and receive cables must be in good condition. They can easily be tested against each other to insure their performance. Bulkhead splices are used to connect the cables under test to the launch and receive cables. Only the highest performance bulkhead splices should be used, and their condition checked regularly, since they are vitally important in obtaining low loss connections.

OPTICAL POWER

The most basic fiber optic measurement is optical power from the end of a fiber. This measurement is the basis for loss measurements as well as the power from a source or at a receiver. Although optical power meters are the primary measurement instrument, OLTSs and OTDRs also measure power differences in testing loss. EIA standard test FOTP-95 covers the measurement of optical power.

Optical power is based on the heating power of the light, and some instruments actually measure the heat when light is absorbed in a detector. While this may work for high-powered lasers, these detectors are not sensitive enough for the power levels typical for fiber optic communication systems. Table 17-2 shows typical power levels in fiber optic systems.

Optical power meters typically use semiconductor detectors since they are extremely sensitive to light in the wavelengths common to fiber optics (Table 17-3). Most fiber optic power meters are available with a choice of three different detec-

Table 17-2 Optical Power Levels of Fiber Optic Communication Systems

Network Type	Wavelength (nm)	Power Range (dBm)	Power Range (W)
Telecom	1310, 1550	+3 to −45	50 nW to 2mW
Datacom	665, 790, 850, 1300	−10 to −30	1 to 100uW
CATV	1310, 1550	+10 to −6	250uW to 10mW

Table 17-3 Characteristics of Detectors Used in Fiber Optic Power Meters

Detector Type	Wavelength Range (nm)	Power Range (dBm)	Comments
Silicon	400–1100	+10 to −70	
Germanium	800–1600	+10 to −60	−70 with small area detectors, +30 with attenuator windows
Indium-Gallium-Arsenide	800–1600	+10 to −70	Small area detectors may overload at high power (>.0 dBm)

tors, silicon (Si), Germanium (Ge), or Indium-Gallium-Arsenide (InGaAs). Silicon photodiodes are sensitive to light in the range of 400 to 1000 nm and germanium and indium-gallium-arsenide photodiodes are sensitive to light in the range of 800 to 1600 nm.

Calibration

Calibrating fiber optic power measurement equipment requires setting up a reference standard traceable to national standards laboratories such as the NIST. Fiber optic power meters have an uncertainty of calibration of about ±5%, compared to NIST primary standards. Limitations in the uncertainty are the inherent inconsistencies in optical coupling, about 1 percent at every transfer, and slight variations in wavelength calibration. NIST is working continuously with instrument manufacturers and private calibration labs to try to reduce the uncertainty of these calibrations.

NIST offers fiber optic power calibration services at 850-nm, 1300-nm, and 1550-nm wavelengths, so most fiber optic power meters offer calibrations at those wavelengths. Fiber optic networks may work at slightly different wavelengths than those calibration wavelengths. For example multimode LED networks use LEDs that are referred to as 1300 nm but have broad spectral outputs, and singlemode networks use lasers referred to as 1310-nm wavelength but actually vary between 1290 and 1330 nm. Since the difference in calibration between 1300 and 1310 nm is insignificant and the actual devices vary from that wavelength significantly, measurements are made only to those calibration wavelengths. Networks using 790-nm transmitters are usually tested at 850-nm calibration, and plastic optical fiber is tested with meters calibrated at 650 nm traceable to other NIST optical power standards.

Recalibration of instruments should be done annually; however, experience has shown that the accuracy of meters rarely changes significantly during that period, as long as the electronics of the meter do not fail. Unfortunately, the calibration of fiber optic power meters requires considerable investment in capital equipment and continual updating of the transfer standards, so very few private calibration labs exist today. Most meters must be returned to the original manufacturer for calibration.

Instrument Resolution versus Measurement Uncertainty

The uncertainty of optical power measurements is about 0.2 dB (5 percent). Loss measurements are likely to have uncertainties of 0.5 dB or more, and optical return loss measurements have about 1 dB uncertainty. Instruments with readouts with a resolution of 0.01 dB are generally only appropriate for laboratory measurements of very low losses such as connectors or splices under 1 dB or for monitoring small changes in loss of power over environmental changes.

Field instruments are better when the instrument resolution is limited to 0.1 dB, since the readings will be less likely to be unstable when being read and more indicative of the measurement uncertainty. This is especially important since field personnel are usually not as well trained in the nuances of measurement uncertainty.

OPTICAL FIBER TESTING

The installer rarely tests fiber or cable before it is installed and terminated except to perform a continuity test before installation to insure no damage has been done to the cable during shipment to the work site. Manufacturers of fiber and cable have already tested the fibers thoroughly and usually provide extensive test data along with the cable.

Continuity Testing

Continuity testing is the most fundamental fiber optic test. It is usually performed with a visible light source, which can be an incandescent light bulb, HeNe laser at 633 nm, or a LED or diode laser at 650 nm, readily seen by the eye. HeNe laser instruments are usually tuned to an output power level of just less than 1 mW, making them Class II lasers, which do not have enough power to harm the eye,

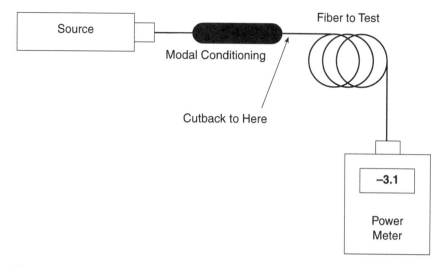

Figure 17-8   Fiber attenuation by cutback method.

but do have enough power to be seen easily over about 4 kilometers of fiber, and even find fiber microbends or breaks by viewing the light shining from the break through the yellow or orange jacket used on most single fiber cables.

Testing Fiber Attenuation

Measuring the fiber attenuation coefficient requires transmitting light of a known wavelength through the fiber and measuring the changes over distance. The conventional method, known as the cutback method (Figure 17-8), involves coupling fiber to the source and measuring the power out of the far end. The fiber is then cut near the source and power measured again. By knowing the power at the source and end of the fiber and the length of the fiber, its attenuation coefficient can be determined by calculating:

$$\text{attenuation coefficient (dB)} = \frac{(P_{end} - P_{source})\ (dB)}{\text{length(km)}}$$

An alternative method of testing fiber, which may be easier in field measurements, involves attaching a fiber pigtail to the source that has a connector on one end and a temporary splice on the other end, similar to the loss measurement of terminated cables. This method introduces more uncertainty in the measurement because of the loss of the splice coupled to the fiber under test, since it may not be easy to accurately calibrate the output power of the pigtail. The best method is to use a bare fiber adapter on the power meter to measure the output of the bare fiber, then attach the splice. Alternately, have the splice attached on the pigtail and couple a large core fiber to the pigtail with the splice and measure the power. The large core fiber will minimize losses in the splice for accurate calibration.

Sources for Loss Measurements

For loss measurements the source can be a fixed-wavelength LED or laser, whichever is appropriate to the fiber being tested. Most multimode fiber systems use LED sources, whereas singlemode fiber systems use laser sources. Thus, testing each of these fibers should be done with the appropriate source. Lasers generally should not be used with multimode fiber, since coherent sources such as lasers have high measurement uncertainties in multimode fiber caused by modal noise. Networks like gigabit Ethernet are too fast for LED sources, so they use lasers and special launch conditions to reduce modal noise. The cable may show significantly lower loss with a laser source than an LED due to different modal conditions. In this circumstance, testing with a source similar to a system source will give more accurate results.

LEDs can be used to test short singlemode cables. However, the wide spectral width of LEDs sometimes overlaps the singlemode fiber cutoff wavelength (the lowest wavelength where the fiber supports only one mode) at lower wavelengths and the 1400-nm hydroxide radical (/OH+): absorption band at the upper wavelengths, creating errors in loss measurements on longer singlemode cables (over about 5 km).

Modal Effects on Attenuation

In order to test multimode fiber optic cables accurately and reproducibly, it is necessary to understand modal distribution, mode control, and attenuation correction factors. Modal distribution in multimode fiber is very important to measurement reproducibility and accuracy. For most field tests, using a source similar to the system source will minimize errors.

Mode Conditioners

There are three basic "gadgets" to condition the modal distribution in multimode fibers (Figure 17-9): mode strippers that remove unwanted cladding mode light, mode scramblers that mix modes to equalize power in all the modes, and mode filters that remove the higher order modes to simulate equilibrium modal distribution (EMD) or steady state conditions. These devices have been described in detail in many articles on testing but are not commonly used in field measurements today, due to the standardization of most link components.

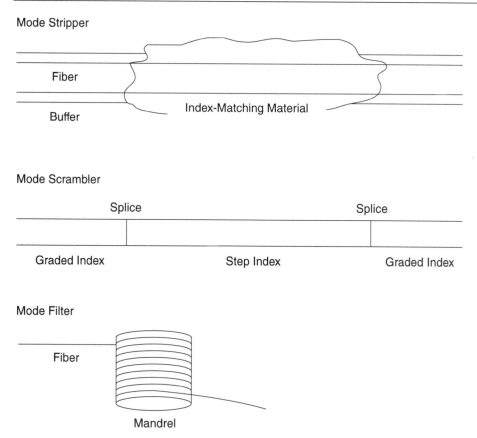

Mode Stripper

Fiber

Buffer

Index-Matching Material

Mode Scrambler

Splice   Splice

Graded Index   Step Index   Graded Index

Mode Filter

Fiber

Mandrel

Figure 17-9   Mode conditioners for multimode graded-index fibers.

Modal Effects in Testing Singlemode Fiber

Testing singlemode fiber is easy compared to multimode fiber. Singlemode fiber, as the name says, only supports one mode of transmission for wavelengths greater than the cutoff wavelength of the fiber. Thus, most problems associated with mode power distribution are no longer a factor. However, it takes a short distance for singlemode fiber to really be singlemode, since several modes may be supported for a short distance after connectors, splices, or sources. Singlemode fibers shorter than 10 meters may have several modes. To insure short cables have only one mode of propagation, one can use a simple mode filter made from a 4- to-6 inch loop of the cable.

Bending Losses

Fiber and cable are subject to additional losses as a result of stress. In fact, fiber makes a very good stress sensor. However, this is an additional source of uncertainty when making attenuation measurements. It is mandatory to minimize stress and/or stress changes on the fiber when making measurements. If the fiber or cable is spooled, it will have higher loss when spooled tightly. It may be advisable to unspool it and respool with less tension. Unspooled fiber should be carefully placed on a bench and taped down to prevent movement. Above all, be careful about how connectorized fiber is placed. Dangling fibers that stress the back of the connector will have significant losses.

Transmission versus OTDR Tests

So far, we have only discussed testing attenuation by transmission of light from a source, but one can also imply fiber losses by backscattered light from a source using an OTDR.

OTDRs are widely used for testing fiber optic cables. Among the common uses are measuring the length of fibers, and finding faults in fibers, breaks in cables, attenuation of fibers, and losses in splices and connectors. They are also used to optimize splices by monitoring splice loss. One of their biggest advantages is that they produce a picture (called a trace) of the cable being tested. Although OTDRs are unquestionably useful for all these tasks, they have error mechanisms that are potentially large, troublesome, and not widely understood.

OTDR Operation

The OTDR (Figure 17-10) uses the lost light scattered in the fiber that is directed back to the source for its operation. It couples a pulse from a high-powered laser source into the fiber through a directional coupler. As the pulse of light passes through the fiber, a small fraction of the light is scattered back toward the source. As it returns to the OTDR, it is directed by the coupler to a very sensitive receiver. The OTDR display (Figure 17-11) shows the intensity of the returned signal in dB as a function of time, converted into distance using the average velocity of light in the glass fiber.

To understand how the OTDR allows measurement, consider what happens to the light pulse it transmits. As it goes down the fiber, the pulse actually "fills" the core of the fiber with light for a distance equal to the pulse width transmitted by the OTDR. In a typical fiber, each nanosecond of pulse width equals about 8 inches (200 mm). Throughout that pulse, light is being scattered, so the longer the pulse width in time, the greater the pulse length in the fiber, the greater will be the amount of backscattered light, in direct proportion to the pulse width. The intensity of the pulse is diminished by the attenuation of the fiber as it proceeds

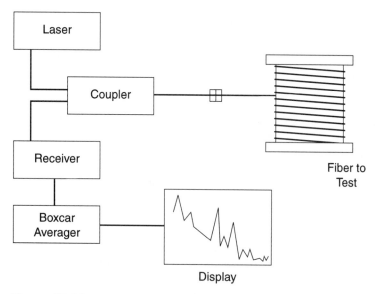

Figure 17-10 An OTDR block diagram.

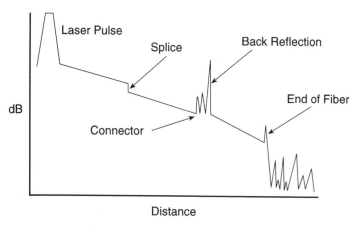

Figure 17-11 OTDR display.

down the fiber, a portion of the pulse's power is scattered back to the OTDR and it is again diminished by the attenuation of the fiber as it returns up the fiber to the OTDR. Thus, the intensity of the signal seen by the OTDR at any point in time is a function of the position of the light pulse in the fiber.

By looking at the reduction in returned signal over time, one can calculate the attenuation coefficient of the fiber being tested. Since the pulse travels out and back, the attenuation of the fiber diminishes the signal in both directions, and the transit time from pulse out to return is twice the one-way travel time. So both the intensity and distance scales must be divided by two to allow for the round-trip path of the light.

If the fiber has a splice or connector, the signal will be diminished as the pulse passes it, so the OTDR sees a reduction in power, indicating the light loss of the joined fibers. If the splice or connector reflects light (see optical return loss), the OTDR will show the reflection as a spike above the backscattered signal. The OTDR can be calibrated to use this spike to measure optical return loss.

The end of the fiber will show as a deterioration of the backscatter signal into noise if it is within the dynamic range of the OTDR. If the end of the fiber is cleaved or polished, one will also see a spike above the backscatter trace. This allows one to measure the total length of the fiber being tested.

In order to enhance the signal to noise ratio of the received signal, the OTDR sends out many pulses and averages the returned signals. And to get to longer distances, the power in the transmitted pulse is increased by widening the pulse width. The longer pulse width fills a longer distance in the fiber as noted earlier. This longer pulse width masks all details within the length of the pulse, increasing the minimum distance between features resolvable with the OTDR.

OTDR Measurement Uncertainties

With the OTDR, one can measure loss and distance. To use them effectively, it is necessary to understand their measurement limitations. The OTDR's distance resolution is limited by the transmitted pulse width. As the OTDR sends out its pulse, crosstalk in the coupler inside the instrument and reflections from the first connector will saturate the receiver. The receiver will take some time to recover, causing a nonlinearity in the baseline of the display. It may take 100 to 1,000 meters before the receiver recovers. It is common to use a long fiber cable called a pulse suppresser between the OTDR and the cables being tested to allow the receiver to recover completely.

The OTDR also is limited in its ability to resolve two closely spaced features by the pulse width. Long distance OTDRs may have a minimum resolution of 250 to 500 meters, while short range OTDRs can resolve features 5 to 10 meters apart. This limitation makes it difficult to find problems inside a building, where distances are short. A visual fault locator is generally used to assist the OTDR in this situation.

When measuring distance, the OTDR has two major sources of error not associated with the instrument itself: the velocity of the light pulse in the fiber and the amount of fiber in the cable. The velocity of the pulse down the fiber is a function of the average index of refraction of the glass. While this is fairly constant for most fiber types, it can vary by a few percent. When making cable, it is necessary to have some excess fiber in the cable, to allow the cable to stretch when pulled without stressing the fiber. This excess fiber is usually 1 to 2 percent. Since the OTDR measures the length of the fiber, not the cable, it is necessary to subtract 1 to 2 percent from the measured length to get the likely cable length. This is very important if one is using the OTDR to find a fault in an installed cable, to keep from looking too far away from the OTDR to find the problem. This variable adds up to 10 to 20 meters per kilometer, therefore it is not ignorable.

When making loss measurements, two major questions arise with OTDR measurement anomalies: why OTDR measurements differ from an OLTS, which tests the fiber in the same configuration in which it is used, and why measurements from OTDRs vary so much when measured in opposite directions on the same splice. And why one direction sometimes shows a gain, not a loss.

In order to understand the problem, it is necessary to consider again how OTDRs work (Figure 17-12). They send a powerful laser pulse down the fiber, which suffers attenuation as it proceeds. At every point on the fiber, part of the light is scattered back up the fiber. The backscattered light is then attenuated by the fiber again, until it returns to the OTDR and is measured.

Note that three factors affect the measured signal: attenuation outbound, scattering, and attenuation inbound.

It is commonly assumed that the backscatter coefficient is a constant, and therefore the OTDR can be calibrated to read attenuation. The backscatter coefficient is, in fact, a function of the core diameter of the fiber (or mode field diameter in singlemode fiber) and the material composition of the fiber (which determines attenuation). Thus, a fiber with either higher attenuation or larger core size will produce a larger backscatter signal.

Accurate OTDR attenuation measurements depend on having a constant backscatter coefficient. Unfortunately, this often is not the case. Fibers with tapers in core size are common, or variations in diameter as the result of variations in pulling speed as the fiber is being made. A small change in diameter (1 percent) causes a larger change in cross-sectional area that directly affects the scattering coefficient and can cause a large change in attenuation (on the order of 0.1 dB). Thus, fiber attenuation measured by OTDRs may be nonlinear along the fiber and produce significantly different losses in opposite directions.

The first indication of OTDR problems for most users occurs when looking at a splice and a gain is seen at the splice. Common sense tells us that passive fibers and splices cannot create light, so another phenomenon must be at work. In fact, a "gainer" is an indication of the difference of backscatter coefficients in the two fibers being spliced.

If an OTDR is used to measure the loss of a splice and the two fibers are identical, the loss will be correct, since the scattering coefficient is the same for

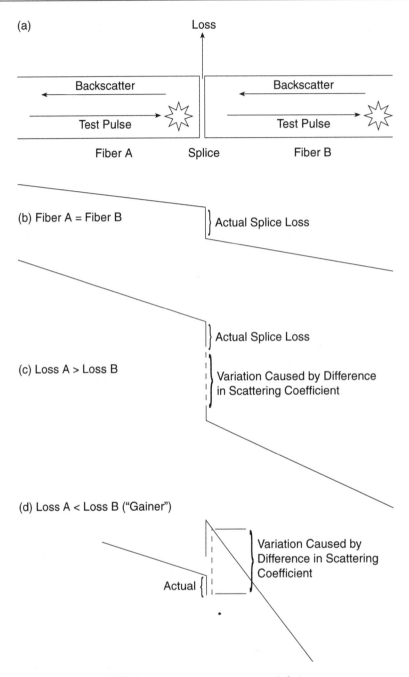

Figure 17-12 OTDR loss measurement uncertainty.

both fibers. This is exactly what you see when breaking and splicing the same fiber, the normal way OTDRs are demonstrated.

If the receiving fiber has a lower backscatter coefficient than the fiber before the splice, the amount of light sent back to the OTDR will decrease after the splice, causing the OTDR to indicate a larger splice loss than actual.

If one looks at this splice in the opposite direction, the effect will be reversed. The amount of backscattered light will be larger after the splice and the loss shown on the OTDR will be less than the actual splice loss. If this increase is larger than the loss in the splice, the OTDR will show a gain at the splice, an obvious error. As many as one-third of all splices will show a gain in one direction.

The usual recommendation is to test with the OTDR in both directions and average the reading, which has been shown to give measurements accurate to about 0.01 dB. But this negates the most useful feature of the OTDR, the ability to work from only one end of the fiber. And all network specifications call for loss testing with a source and meter, which must be done also.

Bandwidth Testing

Fiber's information transmission capacity is limited by two separate components of dispersion: modal and chromatic. Modal dispersion occurs in step-index multimode fiber where the paths of different modes are of varying lengths. Modal dispersion also comes from the fact that the index profile of graded-index (GI) multimode fiber is not perfect.

The second factor in fiber bandwidth is chromatic dispersion. Remember, a prism spreads out the spectrum of incident light, since the light travels at different speeds according to its color and is therefore refracted at different angles. The usual way of stating this is the index of refraction of the glass is wavelength dependent. Thus, a carefully manufactured graded-index multimode fiber can only be optimized for a single wavelength, usually near 1300 nm, and light of other colors will suffer from chromatic dispersion. Even light in the same mode will be dispersed if it is of different wavelengths.

Chromatic dispersion is a bigger problem with LEDs, which have broad spectral outputs, unlike lasers that concentrate most of their light in a narrow spectral range. Chromatic dispersion occurs with LEDs because much of the power is away from the zero dispersion wavelength of the fiber. High speed systems such as FDDI, based on broad output surface emitter LEDs, suffer such intense chromatic dispersion that transmission over only 2 to 3 km of 62.5/125 fiber can be risky.

Modal dispersion is the most commonly tested bandwidth factor. Testing is done by using a narrow spectral width laser source and high-speed receiver to determine dynamic characteristics. Testing can be done by sweeping the frequency of a sine wave and looking for attenuation in the pulse peak height, which leads to a specification of bandwidth at the 3 dB loss point, that is, pulse height is 0.5 the value at low frequency. The alternate method is to measure degradation of pulse risetime.

Chromatic dispersion requires comparing pulse transit times or phase shift as a function of wavelength. Thus, sources of several wavelengths are used and variations in time allow calculating dispersion as a function of wavelength. Although it seems that this could be done with a broad spectral width source such as an LED, the removal of the effects of the spectral characteristics of the LED is very complicated mathematically and every LED is unique in its spectral characteristics, making calibration of test equipment very difficult.

Since all this test equipment must work in the GHz range, it is very expensive. Fortunately, fiber bandwidth characteristics have been very well modeled and the characteristics calculated with precision comparable to actual measurements. There have been at least two models described in detail and one available commercially. The one available commercially (Fotec's Cable Characterizer) calculates bandwidth for multimode fibers based on inputs of fiber modal bandwidth and length and source wavelength and spectral width.

Using the models one can easily determine if the installed fiber is adequate for higher-speed networks such as FDDI. They can help designers design networks with adequate bandwidth for high-speed networks without spending too much on overspecified fiber, and provide a way for the installer or end user to certify cable plants for FDDI and other high-speed networks.

CONNECTOR AND SPLICE LOSS TESTING

Connector loss is the major cause of loss in short fiber optic cable plant runs, making it a very important measurement. Most cable testing is done after the connectors are installed, so the loss includes the connectors. Splice testing is more complicated. Expensive fusion splicers estimate the splice loss themselves, and the data is quite reliable. Splice testing is generally done with an OTDR in both directions and averaged, but the cost and difficulty of splice testing is such that it usually is not done unless the splice quality is suspected from the total end-to-end loss of the cable.

Connectors and splices are tested by the manufacturer to establish a typical loss that can be expected by the experienced user. However, the actual loss of a connector or splice is primarily a function of the installation process, not the component itself. The connector itself is made very precisely, so the loss is determined by how well it is assembled and polished. Splice loss also depends on the skill of the installer, although much less so than on the connector, since polishing is not needed. Large variations in loss from the manufacturer's specifications for loss may mean that the installer needs to improve either the installation process or the test process.

In order to establish a typical loss for connectors or splices, it is necessary to test them in a standardized fashion to allow for comparisons among various connectors. Measurements of connector or splice losses are performed by measuring the transmitted power of a short length of cable and then inserting a connector pair or splice into the fiber (Figure 17-13). This test (designated FOTP-34 by the EIA) can be used for both multimode and singlemode fiber, but the results for multimode fiber are dependent on mode power distribution.

FOTP-34 has three options in modal distribution: (1) EMD or steady state, (2) fully filled, and (3) any other conditions as long as they are specified. Besides mode power distribution factors, the uncertainty of the measured loss is a combination of inherent fiber geometry variations, installed connector characteristics, and the effects of the splice bushing used to align the two connectors.

This test is repeated hundreds or thousands of times by each connector manufacturer to produce data that is quoted as an average value for the connector. This shows the repeatability of their connector design, a critical factor in figuring margins for installations using many connectors. Thus, loss is not the only criterion for a good connector—it must be repeatable, so its average loss can be used for these margin calculations with some degree of confidence.

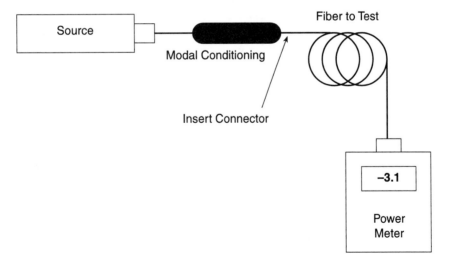

Figure 17-13   Connector insertion loss test.

Inspecting Connectors

Inspecting connectors in operating fiber optic links can be very dangerous if high power levels are present, for instance from a laser source. Whenever inspecting connectors in an installed or operating system, always check the connector with a power meter to ensure no power is present.

Visual inspection of the end surface of a connector with a microscope is one of the best ways to determine the quality of the termination procedure and diagnose problems. A well-made connector will have a smooth, polished, scratch-free finish, and the fiber will not show any signs of cracks or pistoning (where the fiber is either protruding from the end of the ferrule or pulling back into it).

The proper magnification for viewing connectors can be 30 to 400 power. Lower magnification, typical with a jeweler's loupe or pocket magnifier, will not provide adequate resolution for judging the finish on the connector. Too high a magnification tends to make small, ignorable faults look worse than they really are. A better solution is to use medium magnification, but inspect the connector three ways: viewing directly at the end of the polished surface with side lighting, viewing directly with side lighting and light transmitted through the core, and viewing at an angle with lighting from the opposite angle (Figure 17-14).

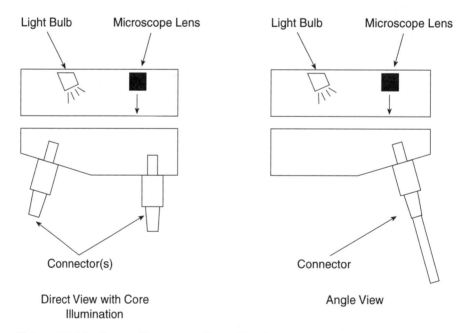

Figure 17-14 Inspecting connection with microscope.

Viewing directly with side lighting allows one to determine if the ferrule hole is of the proper size, the fiber is centered in the hole, and a proper amount of adhesive has been applied. Only the largest scratches will be visible this way, however. Adding light transmitted through the core will make cracks in the end of the fiber, caused by pressure or heat during the polish process, visible.

Viewing the end of the connector at an angle, while lighting it from the opposite side at approximately the same angle, will allow the best inspection for the quality of polish and possible scratches. The shadowing effect of angular viewing enhances the contrast of scratches against the mirror-smooth polished surface of the glass.

One needs to be careful in inspecting connectors, however. The tendency is to be overly critical, especially at high magnification. Only defects over the fiber core are a problem. Chipping of the glass around the outside of the cladding is not unusual and will have no effect on the ability of the connector to couple light in the core. Likewise, scratches only on the cladding will not cause any loss problems.

An alternative way of viewing connector end faces is an interferometer. The interferometer uses a special technique to show a profile of the end of the connector that can determine its flatness or proper curvature for physical contact (PC) connectors. Interferometers are important tools to use for critical connectors such as PC singlemode types, but their size and cost limit their use to manufacturing facilities.

Optical Return Loss in Connectors

If you have ever looked at a fiber optic connector on an OTDR, you are familiar with the characteristic spike that shows where the connector is. That spike is a measure of the back reflection of optical return loss (ORL) of the connector, or the amount of light that is reflected back up the fiber by light reflecting off the interface of the polished end surface of the connector and air. It is called fresnel reflection and is caused by the light going through the change in index of refraction at the interface between the fiber ($n = 1.5$) and air ($n = 1$).

For most systems, that return spike is just one component of the connector's loss, representing about 0.3 dB loss (two air/glass interfaces at 4 percent reflection each), the minimum loss for noncontacting connectors without index-matching fluid. But in high bit-rate singlemode systems, that reflection can be a major source of bit error-rate problems. The reflected light interferes with the laser diode chip, causes mode-hopping, and can be a source of noise. Minimizing the light reflected back into the laser is necessary to get maximum performance out of high bit-rate laser systems, especially the analog modulation (AM) modulated CATV systems.

State-of-the-art connectors will have a return loss of about 40 to 60 dB. Measuring this back reflection can be done in the field with most of today's OTDRs or using a source and power meter per standard test procedure EIA FOTP-107 in the manufacturing process. Generally, ORL is not measured in the field.

Connectorized Cable Testing

After connectors are added to a cable, testing must include the loss of the fiber in the cable plus the loss of the connectors. This is the test that is most often performed in the field after cable installation and termination. On very short cable assemblies (up to 10 m long), the loss of the connectors will be the only relevant loss, while fiber will contribute to the overall losses in longer cable assemblies. In an installed cable plant, one must test the entire cable from end to end, including every component in it, such as splices, couplers, and connectors intermediate patch panels.

In testing connectorized cables, one uses a source with a launch cable attached to calibrate the power being launched into the cable under test and a meter to measure the loss. This test, FOTP-171, was developed along the lines of FOTP-34 for connectors (Figure 17-15). One begins by attaching to the source a launch cable made from the same size fiber and connector type as the cables to be tested. The power from the end of this launch cable is measured by a power meter to calibrate the launch power for the test. Then the cable to test is attached and power measured at the end again. One can calculate the loss incurred in the connectors mating to the launch cable and in the fiber in the cable itself.

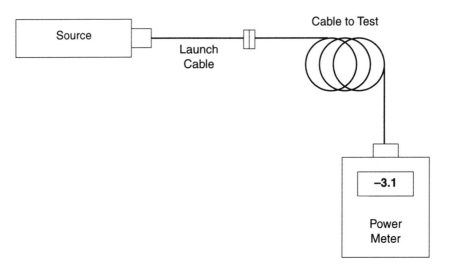

Figure 17-15   Basic fiber optic cable loss test.

Since this only measures the loss in the connector mated to the launch cable, one can add a second cable at the power meter end, called a receive cable, so the cable to test is between the launch and receive cables. Then one measures the loss at both connectors and in everything in between. This is commonly called a double-ended loss test (Figure 17-16).

To obtain accurate loss measurements, it is important to calibrate the launched power from the test source correctly. There have been two interpretations of the calibration of the output of the source in this test. One interpretation (the incorrect one) is that one attaches the launch cable to the source and the receive cable to the meter. The two are then mated and this becomes the "0 dB" reference. The second method only attaches the launch cable to the source and measures the power from the launch cable with the power meter.

With the first method, usually called the "two-cable reference," one has two new measurement uncertainties. First, this method underestimates the loss of the cable plant by the loss of one connection, since that is zeroed out in the calibration process. Secondly, if one has a bad connector on one or both of the test cables, it is hidden by the calibration, since even if the two connectors have a loss of 10 dB, it is not seen by the calibration method used. This can lead to large measurement errors, where losses measured will be higher than actual losses.

In the correct "single-cable" method, the launch power from the cable attached to the source is measured directly by the power meter. This also allows one to measure both connectors on the test cable, since power is referenced to the output power of the launch connector. In addition, one can test the mating quality

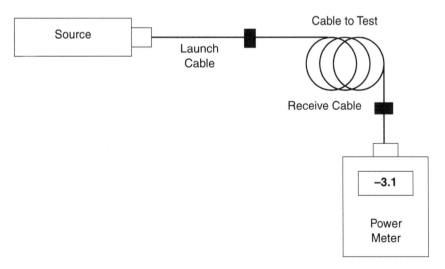

Figure 17-16   Double-ended cable loss test.

of the test cables' connectors by attaching the receive jumper to the meter and then measuring the loss of the connection between the launch and receive jumpers. If this loss is high, one knows there is a problem with the test connectors that must be fixed before actual cable loss measurements should be made.

Obviously, the second method is the proper method. Both methods are detailed in OFSTP-14, the extension of FOTP-171 to include installed cable plants, which also discusses the problems associated with mode power distribution. Also, the loss specifications for the cable plant in all network specifications are written to require that the single-cable launch power calibration method be used.

Finding Bad Connectors

If a test shows a jumper cable to have high loss, there are several ways to find the problem. If you have a microscope, inspect the connectors for obvious defects such as scratches, cracks, or surface contamination. If they look okay, clean them before retesting. Retest the launch cable to make certain it is good. Then retest the jumper cable with the single-ended method, using only a launch cable. Test the cable in both directions. The cable should have higher loss when tested with the bad connector attached to the launch cable, since the large area detector of the power meter will not be affected as much by the typical loss factors of connectors.

Measuring Installed Splice Loss

Most fusion splicers have built-in equipment to inject and detect light transmitted through the splice being made for estimating splice loss. These machines do not require any other means of measuring splice loss. However, for other splice types, it may be desirable to measure splice loss directly.

An OTDR can be used to measure splice loss, but its measurement uncertainty caused by the different characteristics of the two different fibers used make it more of a relative loss measurement than an absolute loss value. However, if knowing the absolute loss of a splice is necessary, measure it with an OTDR in both directions and average the results.

One can also measure the loss of a splice using a technique similar to the FOTP-171 test for connectors. In order to measure the output to the launch fiber (the one connected to the transmitter or test source), one must use a bare fiber adapter on the power meter. After cleaving the fiber, measure the power with the meter and use that as a reference. Once the fiber has been spliced, measuring the loss, including the loss of the length of cable spliced on, can be done at the far end of the fiber being spliced, which may be miles away. Since many splices are usually done at once, moving the power meter to the remote location each time is impractical, so one needs another person with a calibrated meter at the remote location to measure the power and report back the reading to allow calculation of the loss.

Mode Power Distribution Effects on Loss in Multimode Fiber Cables

The biggest factor in the uncertainty of multimode cable loss tests is the mode power distribution caused by the test source. When testing a simple 1 m cable assembly, variations in sources can cause 0.3 to 1 dB variations in measured loss. The effect is similar to the effect on fiber loss discussed earlier, since the concentration of light in the lower order modes as a result of EMD or mode filtering will minimize the effects of gap, offset, and angularity on mating loss by effectively reducing the fiber core size and numerical aperture.

Although one can make mode scramblers and filters to control mode power distribution when testing in the laboratory, it is more difficult to use these in the field. The best way to get reliable measurements is to insure the test source uses a source similar to a system source, not a controlled or restricted launch power distribution. An alternative technique is to use a special mode conditioning cable between the source and launch cable that induces the proper mode power distribution. This can be done with a step-index fiber with a restricted numerical aperture. Experiments with such a cable used between the source have been shown to greatly reduce the variations in mode power distributions between sources. This technique works well with both lab tests of connector loss and field tests of loss in the installed cable plant.

Choosing a Launch Cable for Testing

Obviously, the quality of the launch cable will affect measurements of loss in cable assemblies tested against it. Good connectors with proper polish are obviously needed, but experiments have shown that one cannot improve measurements by specifying tighter specifications on the fiber and connectors. If the fiber is closer to nominal specifications and the connector ferrule is tightly toleranced, one should expect more repeatable measurements, but there are so many variables in the termination process that specifying special parts does not lead to better measurements. Therefore, it is recommended that launch cables be chosen for low loss, but not specified with tighter tolerances in the fiber or connector characteristics.

It is much more important to handle the test cables carefully and inspect the end surfaces of the ferrules in a microscope for dirt and wear or scratches regularly. Remember, the splice bushing used in testing wears out also. Do not use splice bushings with molded plastic alignment sleeves for testing as some wear fast and contaminate the ends of connectors. Use only adapters with metal or ceramic alignment sleeves for test purposes.

Losses from Mismatched Fibers

Fiber mismatches occur for two reasons: the occasional need to interconnect two dissimilar fibers and production variances in fibers of the same nominal dimensions. With two multimode fibers in use today (62.5 and 50 micron cores) and two others that have been used occasionally in the past, connecting dissimilar fibers or using systems designed for one fiber or another is sometimes necessary. Some system manufacturers provide guidelines on using various fibers; some do not. If you connect a smaller fiber to a larger one, the coupling losses will be minimal, often only the fresnel loss (about 0.3 dB). But connecting larger fibers to smaller ones results in substantial losses, not only due to the smaller core size, but also the smaller NA of most small-core fibers.

In the Table 17-4, we show the losses incurred in connecting mismatched fibers. The range of values results from the variability of modal conditions. If the transmitting fiber is overfilled or nearer the source, the loss will be higher. If the fiber is near steady state conditions, the loss will be nearer the lower value.

If you are connecting fiber directly to a source, the variation in power will be approximately the same as for fiber mismatch, except that replacing the smaller fiber with a larger fiber will result in a gain in power roughly equal to the loss in power in coupling from the larger fiber to the smaller one.

Whenever you use a different (and often unspecified) fiber with a system, be aware of differences in fiber bandwidths also. A system may work on paper, with enough power available, but the fiber could have insufficient bandwidth.

TESTING THE INSTALLED FIBER OPTIC CABLE PLANT

The process of testing any fiber optic cable plant during and after installation includes all the procedures covered so far. To thoroughly test the cable plant, one needs to perform three tests—before installation, on each installed segment, and complete end-to-end loss. Practical testing, however, usually means continuity testing each cable before installation to insure there has been no damage to the cable in transit and each segment as it is terminated. Then the entire cable run is tested for end-to-end loss.

One should test the cable on the reel for continuity before installing it to insure no damage was done in shipment from the manufacturer to the job site. Since the cost of installation usually is high, often higher than the cost of materials, it only makes sense to insure that one does not install bad cable. It is generally sufficient just to test continuity, since most fiber is installed without

Table 17.4 Mismatched Fiber Connection Losses (Excess Loss in dB)

| | Transmitting Fiber | | |
Receiving Fiber	62.5/125	85/125	100/140
50/125	0.9–1.6	3.0–4.6	4.7–9
62.5/125	—	0.9	2.1–4.1
85/125	—	—	0.9–1.4

connectors and then terminated in place, and connectors are the most likely problem to be uncovered by testing for loss. After installation and termination, each segment of the cable plant should be tested individually as it is installed, to insure each connector and cable is good. Finally, each end-to-end run (from equipment placed on the cable plant to equipment) should be tested as a final check.

Testing the complete cable plant is done in accordance with another standard test procedure, OFSTP-14 (Figure 17-17). This procedure covers the peculiarities of multimode fiber in detail. In fact, it was written for multimode cables to cover the problems of controlling mode power distribution, but the same procedures apply for singlemode fiber, less the concerns expressed for mode power distribution errors.

For multimode fibers, testing is now usually done at both 850 and 1300 nm, using LED sources. This will prove the performance of the cable for every datacom system, including FDDI and ESCON, and meet the requirements of all network vendors. For singlemode fiber cables, testing is usually done at 1300 nm, but 1550 nm is sometimes required also. The 1550-nm testing will show that the cable can support wavelength division multiplexing (WDM) at 1310 and 1550 nm for future service expansion. In addition, 1550-nm testing can show microbending losses that will not be obvious at 1310 nm, since the fibers are much more sensitive to bending losses at 1550 nm.

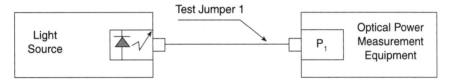

Reference Power Measurement for Method B

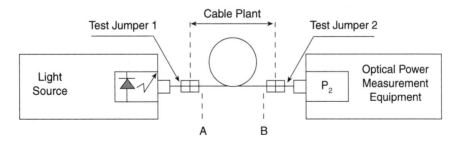

Cable Plant Loss Measurement

Figure 17-17   OFSTP-14 cable plant loss test as required in network specifications.

If cable plant end-to-end loss exceeds total allowable loss, the best solution is to retest each segment of the cable plant separately, checking suspect cables each way, since the most likely problem is a single bad connector or splice. If the cable plant is long enough, an OTDR may be used to find the problem. Bad connectors must then be repolished or replaced to get the loss within acceptable ranges.

What about OTDR Testing?

Once upon a time, OTDRs were used for all testing of installed cable plants. In fact, printouts or pictures of the OTDR traces were kept on record for every fiber in every cable. The power meter and source (or OLTS) have replaced the OTDR for most final qualification testing, since the direct loss test gives a more reliable test of the end-to-end loss than does an OTDR (see OTDR discussion above).

However, the OTDR may need to be used to find bad splices or ORL problems in connectors and splices in a singlemode cable plant. Only with an OTDR can ORL problems be located for correction. Typical back reflection test sets only give a total amount of backscatter or return loss, not the effects of individual components, which is necessary to locate and fix the problem.

The OTDR can also be used to find bad connectors or splices in a high loss cable plant, if the OTDR has high enough resolution to see short, individual cable assemblies. However, if the cables are too short or the splices too near the end of the fiber (as is often the case in pigtails spliced onto singlemode fiber cables), the only way to localize the problem is to use a visual fault locator, preferably a high-powered HeNe laser type, which can shine through the jacket of typical yellow or orange polyvinyl chloride- (PVC) jacketed single-fiber cables. This method of fault location is easiest if single-fiber cables use yellow or orange jackets that are more translucent to the laser light.

Handling and Cleaning Procedures

Connectors and cables should be handled with care. Do not bend cables too tightly, especially near the connectors, as sharp bends can break the fibers. Do not drop the connectors, as they can be damaged by a blow to the optical face. Do not pull hard on the connectors themselves, as this may break the fiber in the backshell of the connector or cause pistoning if the bond between the fiber and the connector ferrule is broken.

COUPLERS AND SWITCHES

Some networks use passive couplers or switches to redirect the fiber path. These devices must be tested for loss just as is any other component, although each possible light path needs testing individually. Multimode components will be sensitive to mode power distribution and need to be tested carefully to get accurate results.

Fiber Optic Couplers

Couplers split or combine light in fibers. They may be simple splitters or 2×2 couplers, or up to 64×64 ports star couplers. Most are made by fusing fibers under high temperatures, which causes light to split or combine in appropriate ratios. Relevant specifications for couplers are the coupling ratios of each port or the consistency across all the ports, crosstalk, and the excess loss caused by the fusing. Excess loss is the difference between the sum of all the outputs and the sum of all the inputs. When used in laser-based systems, couplers may need testing for optical return loss and also for wavelength dependence.

Thus, testing couplers involves coupling a test source to each input port in turn and measuring all the outputs for consistency, then summing all the output powers and subtracting that number from the input power to calculate excess loss. Connectorized couplers are tested like connectorized cables, using a launch cable, whereas couplers with bare fibers must use a cutback method or a pigtail and temporary splice to couple the launch source.

Singlemode couplers have another characteristic that must be considered: They are very wavelength sensitive. Most couplers are optimized at one wavelength unless they are specially designed for both 1310- and 1550-nm operation. Some are even built to be wavelength division multiplexers, coupling light from 1310- and 1550-nm lasers into separate output ports. Therefore, sources for testing couplers must be accurately characterized for wavelength to minimize measurement uncertainty.

Fiber Optic Switches

Switches transfer light from one fiber to another. As with couplers, switch testing involves measuring the loss in the switch by measuring the input from a source and the appropriate output for each switch position. In multimode components, mode power distribution can cause variation in switch losses or coupling ratios.

FIBER OPTIC DATALINKS

Fiber optic transmission systems all work similarly; they contain a transmitter that takes an electrical input and converts it to an optical output from a laser diode or LED. The light from the transmitter is coupled into the fiber with a connector and is transmitted through the fiber optic cable plant. The light is ultimately coupled to a receiver where a detector converts the light into an electrical signal, which is then conditioned properly for use by the receiving equipment. Most networks or datalinks will be bidirectional and full duplex, transmitting and receiving simultaneously. They will have two links as shown in Figure 17-18, operating in opposite directions. Just as with copper wire or radio transmission, the performance of the fiber optic data link can be determined by how well

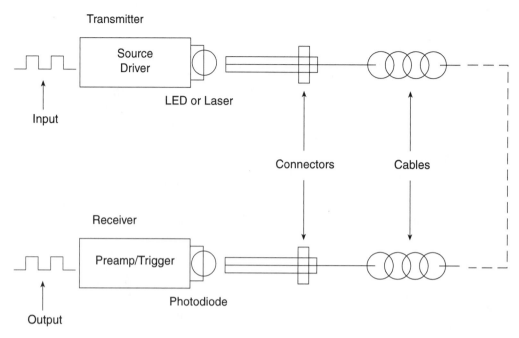

Figure 17-18   Typical fiber optic link.

the reconverted electrical signal out of the receiver matches the input to the transmitter.

The ability of any fiber optic system to transmit data ultimately depends on the optical power at the receiver as shown in Figure 17-19, which shows the datalink bit-error rate as a function of optical power at the receiver. Either too little or too much power will cause high bit-error rates. Too much power, and the receiver amplifier saturates, too little and noise becomes a problem. This receiver power depends on two basic factors: how much power is launched into the fiber by the transmitter and how much is lost by attenuation in the optical fiber cable that connects the transmitter and receiver.

Datalink testing is done with a power meter that measures the optical power first at the receiver and then at the transmitter (with its power coupled into a known good test cable, usually one of the launch cables used for testing the cable plant) as shown in Figure 17-20.

What Goes Wrong on Fiber Optic Installations?

In installing and testing fiber optic networks, the first problem routinely encountered is incorrect fiber optic connections. A fiber optic link consists of two fibers, transmitting in opposite directions, to provide full duplex communications. It is

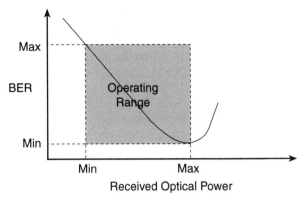

Figure 17-19   Bit-error rate (BER) performance of fiber optic systems.

not uncommon for the transmit and receive fibers to be switched. A visual fiber tracer will make it easy to verify the proper connections quickly. The visual tracer can trace the fiber through the cables, patch panels, and other components to the far end.

If the fibers are connected correctly but the network still does not work, the next thing to check is the receiver power level. If the receiver power level is within specification, the problem is likely in the network electronics. If the receiver power is low, test the transmitter power to see if it is within specifications. If transmitter power is adequate, the cable plant is the problem.

Test the complete cable plant—including all individual jumper or trunk cables—for loss, using a power meter and source and the double-ended method

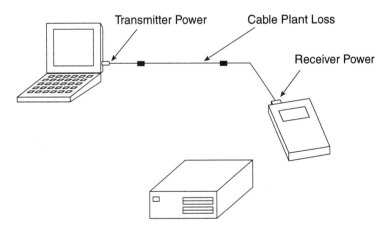

Figure 17-20   Troubleshooting a fiber optic link.

described in the section on testing the installed cable plant. Use the double-ended method, since system margin specifications include the loss of connectors on both ends of the fiber. If the end-to-end (transmitter to receiver) loss measurement for a given fiber is within the network margin specification, the data should be recorded for future reference.

If the loss is too low, notation should be made that that fiber will probably need an inline attenuator to reduce receiver power to acceptable levels. If the loss is too high, it will be necessary to retest each link of the complete cable run to find the bad link.

Possible causes of high end-to-end link loss are bad connectors, bad splice bushings in patch panels, cables bent too tightly around corners, broken fibers in cables, or even bad launch or receive cables or instruments. There are only two ways to find the problem: testing each segment of the cable individually to find the problem and issuing an OTDR if the lengths are long enough for viewing with the limited resolution of the OTDR.

Do not use an OTDR for measuring end-to-end loss. It will not accurately measure actual link loss as seen by the actual transmitters and receivers of the fiber optic link. As normally used, the OTDR will not count the end connectors' loss. The OTDR uses a laser with very restricted mode power distribution, which minimizes the loss of the fiber and the intermediate connectors. Finally, the difference in backscattering coefficients of various fibers leads to imprecise connector loss measurements.

Surviving with Fiber Optics

Once the installation is complete, the cable plant tested, the network equipment running smoothly, what is likely to go wrong in a fiber optic network? Fortunately, not much. One of the biggest selling points for fiber optics has been its reliability. But there are potential problems that can be addressed by the end user.

With the cable plant, the biggest problem is what the telcos call "backhoe fade," where someone mistakenly cuts or breaks the cable. Although this most often happens when an underground cable is dug up, it can happen when an electrician is working on cables inside a building. Outdoors, the best defense is to mark where cables are buried and bury a marker tape above the cable that will, it is hoped, be dug up first. Inside buildings, using orange or yellow jacket cable instead of black or gray will make the fiber cable more visible and distinctive. Outside cable faults are best found by using an OTDR to localize the fault, then having personnel scout the area looking for obvious damage. Inside buildings, the short distances make OTDRs unusable, so a visual fault locator is necessary. Another problem is breaking the cable just behind the connectors in patch panels.

This is a difficult fault to find, but a visual fault locator is often the best way. Unless the jumper cables are quite long, an OTDR will not help at all.

Within the fiber optic link, the most likely component to fail is the LED or laser transmitter, since it is the most highly stressed component in the link. Lasers are feedback stabilized to maintain a constant output power, so they tend to fail all at once. LEDs will drop in power output as they age, but the timeframe is quite long, 100,000 to 1 million hours. If there is no power at the receiver the next place to check should be the transmitter LED or laser, just to isolate the problem to either the transmitter or the cable plant. Receivers are low-stressed devices and highly reliable, but the electronics behind them can fail. If there is receiver power but no communications, a loopback test to see if the receiver is working is the best test of its status.

REVIEW QUESTIONS

1. _____ is the most important parameter and is required for almost all fiber optic tests.
   a. Attenuation
   b. Backscatter
   c. Optical power
   d. Receiver power

2. Match the items on the left with related phrases on the right.
   ____ Power meter   a. detects a 2kH "tone"
   ____ OTDR   b. detects poor splices or cable breaks in
   ____ Fiber identifier   short lengths of fiber
   ____ Visible fault locator   c. measures optical power
   ____ Microscope   d. detects faults in long lengths of fiber
   ____ Attenuator   e. introduces a measured loss into the link
          f. detects scratches and polishing defects
          in connectors

3. Optical fiber is sometimes tested for _____ before it is installed.
   a. back reflection
   b. splice loss
   c. continuity
   d. return loss

4. 0 dBm is equal to _____
   a. 1 mW.
   b. −30 dBµ.
   c. both a and b
   d. neither a nor b

5. When testing for loss, the launch cable must be compatible with the

   _____
   a. source wave length.
   b. cable type.
   c. fiber type (62.5, 50, SM).
   d. length of the cable.

6. An OTDR uses _____ to characterize fibers and find faults.
   a. absorption
   b. attenuation
   c. backscattering
   d. LEDs

7. One advantage an OTDR has over a power meter and source is

   _____
   a. lower price.
   b. more accurate attenuation reading.
   c. it requires access to only one end of the fiber.
   d. no measurement uncertainty.

8. To detect splice loss in a 1.5 km yellow-jacketed SM fiber a
   _____ is used.
   a. OLTS
   b. OTDR
   c. Visual fault locator
   d. OCWR

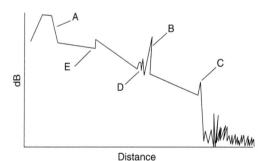

Distance

9. Using the accompanying trace, identify the following:
   1. splice _____
   2. initial pulse _____
   3. connector _____
   4. end of fiber _____
   5. back reflections _____
   6. gain _____

A

GLOSSARY OF FIBER OPTIC TERMS

Absorption: That portion of fiber optic attenuation that is the result of conversion of optical power to heat.

Analog: Signals that are continually changing, as opposed to being digitally encoded.

Attenuation: The reduction in optical power as it passes along a fiber, usually expressed in decibels (dB). See *Optical loss*.

Attenuation coefficient: Characteristic of the attenuation of an optical fiber per unit length, in dB/km.

Attenuator: A device that reduces signal power in a fiber optic link by inducing loss.

Average power: The average over time of a modulated signal.

Back reflection, optical return loss: Light reflected from the cleaved or polished end of a fiber caused by the difference of refractive indices of air and glass. Expressed in dB relative to incident power.

Backscattering: The scattering of light in a fiber back toward the source, used to make OTDR measurements.

Bandwidth: The range of signal frequencies or bit rate within which a fiber optic component, link, or network will operate.

Bending loss, microbending loss: Loss in fiber caused by stress on the fiber bent around a restrictive radius.

Bit: An electrical or optical pulse that carries information.

Bit-error rate (BER): The fraction of data bits transmitted that are received in error.

Buffer: A protective coating applied directly on the fiber.

Cable: One or more fibers enclosed in protective coverings and strength members.

Cable plant, fiber optic: The combination of fiber optic cable sections, connectors, and splices forming the optical path between two terminal devices.

CATV: An abbreviation for community antenna television or cable TV.

Chromatic dispersion: The temporal spreading of a pulse in an optical waveguide caused by the wavelength dependence of the velocities of light.

Cladding: The lower refractive index optical coating over the core of the fiber that traps light into the core.

Connector: A device that provides for a demountable connection between two fibers or a fiber and an active device and provides protection for the fiber.

Core: The center of the optical fiber through which light is transmitted.

Coupler: An optical device that splits or combines light from more than one fiber.

Cutback method: A technique for measuring the loss of bare fiber by measuring the optical power transmitted through a long length then cutting back to the source and measuring the initial coupled power.

Cutoff wavelength: The wavelength beyond which singlemode fiber only supports one mode of propagation.

dB: Optical power referenced to 1 microwatt.

dBm: Optical power referenced to 1 milliwatt.

Decibel (dB): A unit of measurement of optical power that indicates relative power on a logarithmic scale, sometimes called dBr. dB = 10 log (power ratio)

Detector: A photodiode that converts optical signals to electrical signals.

Digital: Signals encoded into discrete bits.

Dispersion: The temporal spreading of a pulse in an optical waveguide. May be caused by modal or chromatic effects.

Edge-emitting diode (E-LED): A LED that emits from the edge of the semiconductor chip, producing higher power and narrower spectral width.

End finish: The quality of the end surface of a fiber prepared for splicing or terminated in a connector.

Equilibrium modal distribution (EMD): Steady state modal distribution in multimode fiber, achieved some distance from the source, where the relative power in the modes becomes stable with increasing distance.

ESCON: IBM™ standard for connecting peripherals to a computer over fiber optics. Acronym for enterprise system connection.

Excess loss: The amount of light lost in a coupler beyond that inherent in the splitting to multiple output fibers.

Fiber Distributed Data Interface (FDDI): 100 Mb/s ring architecture data network.

Ferrule: A precision tube that holds a fiber for alignment for interconnection or termination. A ferrule may be part of a connector or mechanical splice.

Fiber identifier: A device that clamps onto a fiber and couples light from the fiber by bending, to identify the fiber and detect high-speed traffic of an operating link or a 2 kHz tone injected by a test source.

Fiber optics: Light transmission through flexible transmissive fibers for communications or lighting.

Fiber tracer: An instrument that couples visible light into the fiber to allow visual checking of continuity and tracing for correct connections.

FO: Common abbreviation for fiber optic.

Fresnel reflection, back reflection, optical return loss: Light reflected from the cleaved or polished end of a fiber caused by the difference of refractive indices of air and glass. Typically 4 percent of the incident light.

Fusion splicer: An instrument that splices fibers by fusing or welding them, typically by electrical arc.

Graded index (GI): A type of multimode fiber that uses a graded profile of refractive index in the core material to correct for dispersion.

Index matching fluid: A liquid used of refractive index similar to glass used to match the materials at the ends of two fibers to reduce loss and back reflection.

Index profile: The refractive index of a fiber as a function of cross section.

Index of refraction: A measure of the speed of light in a material.

Insertion loss: The loss caused by the insertion of a component such as a splice or connector in an optical fiber.

Jacket: The protective outer coating of the cable.

Jumper cable: A short single-fiber cable with connectors on both ends used for interconnecting other cables or testing.

Laser diode (ILD): A semiconductor device that emits high-powered, coherent light when stimulated by an electrical current. Used in transmitters for single-mode fiber links.

Launch cable: A known good fiber optic jumper cable attached to a source and calibrated for output power used for loss testing. This cable must be made of fiber and connectors of a matching type to the cables to be tested.

Light-emitting diode (LED): A semiconductor device that emits light when stimulated by an electrical current. Used in transmitters for multimode fiber links.

Link, fiber optic: A combination of transmitter, receiver, and fiber optic cable connecting them capable of transmitting data. May be analog or digital.

Long wavelength: A commonly used term for light in the 1300 and 1550 nm ranges.

Loss budget: The amount of power lost in the link. Often used in terms of the maximum amount of loss that can be tolerated by a given link.

Loss, optical: The amount of optical power lost as light is transmitted through fiber, splices, couplers, and the like.

Margin: The additional amount of loss that can be tolerated in a link.

Mechanical splice: A semipermanent connection between two fibers made with an alignment device and index matching fluid or adhesive.

Micron (m): A unit of measure, 10^{-6} m, used to measure wavelength of light.

Microscope, fiber optic inspection: A microscope used to inspect the end surface of a connector for flaws or contamination or a fiber for cleave quality.

Modal dispersion: The temporal spreading of a pulse in an optical waveguide caused by modal effects.

Mode: A single electromagnetic field pattern that travels in fiber.

Mode field diameter: A measure of the core size in singlemode fiber.

Mode filter: A device that removes optical power in higher-order modes in fiber.

Mode scrambler: A device that mixes optical power in fiber to achieve equal power distribution in all modes.

Mode stripper: A device that removes light in the cladding of an optical fiber.

Multimode fiber: A fiber with core diameter much larger than the wavelength of light transmitted that allows many modes of light to propagate. Commonly used with LED sources for lower-speed, short-distance links.

Nanometer (nm): A unit of measure, 10^{-9} m, used to measure the wavelength of light.

Network: A system of cables, hardware, and equipment used for communications.

Numerical aperture (NA): A measure of the light acceptance angle of the fiber.

Optical amplifier: A device that amplifies light without converting it to an electrical signal.

Optical fiber: An optical waveguide, comprised of a light-carrying core and cladding that traps light in the core.

Optical loss test set (OLTS): A measurement instrument for optical loss that includes both a meter and source.

Optical power: The amount of radiant energy per unit time, expressed in linear units of Watts or on a logarithmic scale, in dBm (where 0 dB = 1 mW) or dB (where 0 dB = 1 W).

Optical return loss, back reflection: Light reflected from the cleaved or polished end of a fiber caused by the difference of refractive indices of air and glass. Typically 4 percent of the incident light. Expressed in dB relative to incident power.

Optical switch: A device that routes an optical signal from one or more input ports to one or more output ports.

Optical time domain reflectometer (OTDR): An instrument that uses backscattered light to find faults in optical fiber and infer loss.

Overfilled launch: A condition for launching light into the fiber where the incoming light has a spot size and NA larger than accepted by the fiber, filling all modes in the fiber.

Photodiode: A semiconductor that converts light to an electrical signal, used in fiber optic receivers.

Pigtail: A short length of fiber attached to a fiber optic component such as a laser or coupler.

Plastic-clad silica (PCS) fiber: A fiber made with a glass core and plastic cladding.

Plastic optical fiber (POF): An optical fiber made of plastic.

Power budget: The difference (in dB) between the transmitted optical power (in dBm) and the receiver sensitivity (in dBm).

Power meter, fiber optic: An instrument that measures optical power emanating from the end of a fiber.

Preform: The large diameter glass rod from which fiber is drawn.

Receive cable: A known good fiber optic jumper cable attached to a power meter used for loss testing. This cable must be made of fiber and connectors of a matching type to the cables to be tested.

Receiver: A device containing a photodiode and signal conditioning circuitry that converts light to an electrical signal in fiber optic links.

Refractive index: A property of optical materials that relates to the velocity of light in the material.

Repeater, regenerator: A device that receives a fiber optic signal and regenerates it for retransmission, used in very long fiber optic links.

Scattering: The change of direction of light after striking small particles that causes loss in optical fibers.

Short wavelength: A commonly used term for light in the 665, 790, and 850 nm ranges.

Singlemode fiber: A fiber with a small core, only a few times the wavelength of light transmitted, that allows only one mode of light to propagate. Commonly used with laser sources for high-speed, long-distance links.

Source: A laser diode or LED used to inject an optical signal into fiber.

Splice, fusion or mechanical: A device that provides for a connection between two fibers, typically intended to be permanent.

Splitting ratio: The distribution of power among the output fibers of a coupler.

Steady state modal distribution: Equilibrium modal distribution (EMD) in multimode fiber, achieved some distance from the source, where the relative power in the modes becomes stable with increasing distance.

Step-index fiber: A multimode fiber where the core is all the same index of refraction.

Surface emitter LED: A LED that emits light perpendicular to the semiconductor chip. Most LEDs used in datacommunications are surface emitters.

Talkset, fiber optic: A communication device that allows conversation over unused fibers.

Termination: Preparation of the end of a fiber to allow connection to another fiber or an active device, sometimes also called "connectorization".

Test cable: A short single-fiber jumper cable with connectors on both ends used for testing. This cable must be made of fiber and connectors of a matching type to the cables to be tested.

Test kit: A kit of fiber optic instruments, typically including a power meter, source, and test accessories, used for measuring loss and power.

Test source: A laser diode or LED used to inject an optical signal into fiber for testing loss of the fiber or other components.

Total internal reflection: Confinement of light into the core of a fiber by the reflection off the core-cladding boundary.

Transmitter: A device that includes a LED or laser source and signal conditioning electronics used to inject a signal into fiber.

Visual fault locator: A device that couples visible light into the fiber to allow visual tracing and testing of continuity. Some are bright enough to allow finding breaks in fiber through the cable jacket.

Watts: A linear measure of optical power, usually expressed in milliwatts (mW), microwatts (µW), or nanowatts (nW).

Wavelength: A measure of the color of light, usually expressed in nanometers (nm) or microns (µm).

Wavelength division multiplexing (WDM): A technique of sending signals of several different wavelengths of light into the fiber simultaneously.

Working margin: The difference (in dB) between the power budget and the loss budget (i.e., the excess power margin).

B

FIBER OPTIC STANDARDS

Widespread use of any technology depends on the existence of acceptable standards. Standards must include primary measurement standards, component standards, network standards, standard test methods, and calibration standards. In fiber optics, this means standardized specifications for fiber, cables, connectors, and splices and test procedures for fibers, cables, connectors, and splices under many varying environmental conditions. Primary and transfer standards for optical power, attenuation, bandwidth, and the physical characteristics of fiber are also required.

These standards are developed by various groups working together, all of which are listed with contact information in Appendix C. Network standards come from American National Standards Institute (ANSI), Institute of Electrical and Electronics Engineers (IEEE), International Electrotechnical Commission (IEC), International Organization for Standardization (IOS), Telcordia (formerly Bellcore), and other groups worldwide. The component and testing standards come from some of these same groups, as well as from the Electronic Industries Alliance (EIA) in the United States and internationally from the ISO and IEC, and other groups worldwide. Primary and transfer standards are developed by national standards laboratories such as National Institute of Standards and Technology (NIST), formerly the National Bureau of Standards which exist in almost all countries to regulate all measurement standards. International cooperation is available to ensure worldwide conformance to all absolute standards.

We must also discuss "de facto" standards, those generally accepted standards for components and systems that are widely accepted in the marketplace. In fact, we want to discuss all of those and their status in today's fiber optic systems.

DE FACTO STANDARDS COME FIRST

In any fast developing technology such as fiber optics, there is always resistance to the development of standards. Critics say standards stifle technology development. Some critics object because it is not their standard that is proposed, and in some cases, nobody really knows what standards are best because the technology is still under development. Given these circumstances, users must choose the best solutions for their problems and forge ahead. In fiber optics, those who have gone ahead and committed heavily to the technology or who have marketing strength have established many of today's standards.

In telecom systems, there are many types of systems but all are operating on singlemode fiber at 1310 or 1550-nm wavelengths. Bit rates of 1.544 Mbits/S up to 2.5 Gbits/S are already in operation, with wavelength division multiplexing (WDM) giving much higher rates. Today, Synchronous Optical Network (SONET) in the United States or Synchronous Digital Hierarchy (SDH) in the rest of the world is the network protocol of choice. That appears to be changing as the effect of the Internet drives everyone to Internet protocol (IP) networks.

In datacom systems (the generic category that includes datalinks and local area networks [LANs]), the situation is reaching consensus. Four multimode fibers have been used in datacom systems: 50/125, 62.5/125, 85/125, and 100/140 (core/clad in microns), but 62.5/125 fiber has been dominant. It originally was chosen as the preferred fiber for Fiber Distributed Data Interface (FDDI) and Enterprise System Connection (ESCON), became adopted by all versions of Ethernet, and the U.S. government is using 62.5/125 exclusively in offices (FED STD 1070). Connectors have usually been ST style, but the EIA/TIA 568 Standard calls for the SC. The new small form factor (SFF) connectors are now the multimode connector of choice for the networking equipment manufacturers, as they offer higher density connections and reduce electronics cost.

While short wavelength light-emitting diode (LED) (820-850 nm) systems have been most popular for Ethernet at 10 MB/s, the higher bit rates of faster systems are requiring 1300-nm LEDs due to the limiting effects of chromatic dispersion in the fiber. The development of low-cost 850-nm vertical cavity surface-emitting lasers (VCSELS) operating with multimode fiber has made Gigabit Ethernet possible using lower-cost components, enhancing fiber optics as a networking technology.

INDUSTRY STANDARDS ACTIVITIES

In light of these de facto standards, many groups are working to develop standards that are acceptable throughout the industry.

Primary Standards

The keeper of primary standards in the United States is the Department of Commerce, National Institute of Standards and Technology (NIST). Although some optical standards work is done at Gaithersburg, Maryland, fiber optic and laser activity is centered at Boulder, Colorado. Today, NIST is actively working with all standards bodies to determine the primary reference standards needed and to provide for them. With fiber optics applications, their concern has been with fiber measurements, such as attenuation and bandwidth, mode field diameter for singlemode fiber, and optical power measurements.

NIST standards are in place for fiber attenuation and optical power measurements, the most important measurement in fiber optics. Since all other measurements require measuring power, several years ago NIST ran a "round-robin" that showed up to 3 dB differences (50 percent) in power measurements among participants. An optical power calibration program at NIST has resulted in reliable transfer standards at 850 nm, 1300 nm and 1550 nm. Using new transfer standards, measurements of better than 5 percent accuracy should be easily obtained.

Component and Testing Standards

Several groups are looking at fiber optic testing standards, but the most active by far is the EIA in the United States and the ISO worldwide. EIA FO-6 and FO-2 committees meet at least twice a year to discuss technical issues and review progress on the writing of standards test procedures and component specifications. At the current time, there are over 100 EIA fiber optic test procedures (FOTPs) in process or published and many component specifications are being prepared. The EIA should be contacted for an up-to-date list of currently published standards as well as those in process.

In addition to being a standards-writing body, the FO-6 and FO-2 committees are a forum for the discussion of technical issues, relevant to the FOTPs being prepared, and are sometimes scenes of heated debate over these issues. But real progress is being made in defining relevant tests for fiber optic component and system performance.

Within the United States, Bell Communications Research (Bellcore), the spin-off research and development organization for the divested Regional Bell Operating companies (RBOCs), was commissioned to set standards for its RBOCs by issuing technical advisories (TAs) on subjects of mutual interest. Bellcore was sold to a commercial company and renamed Telcordia, thus its impact on future standards is unknown.

Internationally, almost every country has its own standards bodies, but most work through ISO and the IEC to produce mutually acceptable standards. The IEC work is at least as large in scope as the EIA.

System Standards

Most early fiber optic systems were compatible with some electrical standards, such as T-3, RS-232, and so forth, but each manufacturer used its own protocol on the optical part of the network. As a result, there was little compatibility in fiber optic systems. Even in telephony, fiber optic links developed as adapters for standard T-carrier systems, so each manufacturer used its own protocol. Bellcore developed SONET standards and CCITT did SDH to provide a standard protocol for telephony.

Work is now done by ANSI and IEEE on developing standard systems for computer networks. The ANSI FDDI (X3T9.5 committee) is a high bit-rate system for computer networks that has reached commercial reality. Another ANSI committee (X3T9.3) is working on the even faster Fibre Channel specification for GB/s data communications. The IEEE standards include a token-ring LAN (802.5), metropolitan area LAN (802.6), and fiber versions of all varieties of Ethernet (802.3). From the vendor front, ESCON was developed by IBM to connect mainframes to peripherals and has been adopted by the entire IBM-compatible mainframe industry. Asynchronous transfer mode (ATM) is a fiber or copper-based LAN using protocol borrowed from the telephony technology developed as part of SONET and was developed into a LAN technology by an industry group advocating its use. So network standards may be developed by any number of groups who will then control the issues of compatibility and interoperability.

Standards change continuously. To keep informed on the current status of fiber optic standards, contact the standards bodies involved for latest information.

RESOURCE GUIDE TO FIBER OPTICS

To assist you in getting started, we have compiled a list of resources that will help you to obtain the basic information needed.

The Fiber Optic Association is a professional society for all of those involved in fiber optics. They develop training and certification programs.

The Fiber Optic Association, Inc.
Box 230851
Boston, MA 02123-0851
617-469-2FOA
www.TheFOA.org

Textbooks: These are good, basic textbooks on fiber optic technology, recommended for beginners but still good references for the knowledgeable user.

Hayes & Rosenberg, *Data, Voice and Video Cabling*, Delmar

Jeff Hecht, *Understanding Fiber Optics*, Howard W. Sams Books

J. Refi, *Fiber Optic Cable, A Lightguide*, ABC Teletraining

Martin Weik, *Fiber Optic Standard Dictionary*, Van Nostrand Reinhold

Application Notes: These companies have extensive libraries of applications literature. Contact them for a current list of notes available.

Belden (Box 1980, Richmond, IN 47375, 317-983-5200),

Corning (Corning, NY 14831, 607-974-4411),

Fotec (151 Mystic Ave., Medford, MA 02155, 1-800-537-8254, 1-781-396-6155, fax 1-781-396-6395. www.fotec.com)

Sourcebooks/Directories: These are compiled lists of vendors and their products, especially good for locating sources of supply for fiber optic components.

Fiberoptic Product News Buying Guide (Gordon Publications, 13 Emery Ave, Randolph, NJ 07869, 201-361-9060)

The Fiber Optic Yellow Pages (IGI, 214 Harvard Ave., Boston, MA 02134, 617-232-3111)

KMI Directory of Worldwide Fiber Optic Suppliers (KMI, 31 Bridge St. Newport, RI 02840, 401-849-6771)

Laser Focus Buyer's Guide (Penwell Publications, Ten Tara Blvd., Nashua, NH 03062, 603-891-0123)

Lightwave Buyer's Guide (Penwell Publications, above)

Photonics Handbook (Laurin Publishing, Berkshire Common, Pittsfield, MA 01202, 413-499-0514)

Magazines: These are periodical magazines on fiber optics, carrying technical articles and industry news.

Cabling Standards Update, 1989A Santa Rita Rd., Pleasanton, CA 94566 (925-846-9900)

Fiberoptic Product News (Gordon Publications, above)

Lightwave (Penwell Publications, above)

Tradeshows:

ECOC (Ecole Polytechnic Federale, CH-1015 Lausanne, Switzerland, 41-21-693-3338)

Interopto(OITDA, Toranamon 1-Chome Mori Bldg. 1-19-5 Toranamon, Minato-Ku, Tokyo 105, Japan, 81-3-3508-2091)

O-E/Lase (SPIE, Box 10, Bellingham, WA, 206-676-3290)

OFC/OFS (Optical Society of America, 1816 Jefferson Pl. NW, Washington, DC 20036, 202-223-8130)

Standards Groups:

International Organization for Standardization (ISO)
1, rue de Varembé
Case postale 56
CH-1211 Genèva 20
Switzerland
Telephone +41 22 749 0111
Fax +41 22 733 3430
E-mail central@iso.ch
Web http://www.iso.ch

International Electrotechnical Commission (IEC)
3, rue de Varembé
P.O. Box 131
CH - 1211 Genèva 20
Switzerland
Telephone +41 22 919 0211
Fax +41 22 919 0300
E-mail info@iec.ch

American National Standards Institute (ANSI)
11 West 42nd Street
New York, NY 10036
Telephone 212-642-4900
Fax 212.398.0023
Web http://web.ansi.org

Telcordia Technologies (formerly Bellcore)
445 South Street.
Morristown, NJ 07962
Telephone (973) 829-2000
Web http://www.bellcore.com

Electronic Industries Alliance/Telecommunications Industry Association
(EIA/TIA)
2500 Wilson Blvd.
Arlington, VA 22201
Telephone 703-907-7500
Web http://www.eia.org

Institute of Electrical and Electronics Engineers (IEEE)
IEEE Corporate Office
3 Park Avenue, 17th Floor

New York, NY 10016-5997 USA
Telephone 212- 419 -7900
Fax 212 -752 -4929
Web http://www.ieee.org

Most standards can be purchased from:

Global Engineering Documents
15 Inverness Way East
Englewood, CO, 80112 USA
E-Mail global@ihs.com
Telephone +1-800-854-7179 (US)
Telephone +1-303-397-7956 (outside the US)
Fax +1-303-397-2740
Web http://global.ihs.com/

ACRONYMS

AM	Analog modulation
ANSI	American National Standards Institute
APC	Angled physical contact
ATM	Asynchronous transfer mode
BWDP	Bandwidth-distance product
CAD	Computer aided design
CATV	Community antenna television
CEV	Controlled environment vault
CFC	?
CSL	?
CW	Carrier wave
DSL	Digital subscriber loop
DWDM	Dense wavelength division multiplexing
EIA	Electronic Industries Alliance
EMD	Equilibrium modal distribution
EMI	Electromagnetic interference
ENG	Electronic news gathering
ERK	Emergency restoration kit
ESCON	Enterprise System Connection
FDDI	Fiber Distributed Data Interface
FOTP	Fiber optic test procedures
FTTC	Fiber to the curb
FTTH	Fiber to the home
HIPPI	High performance parallel interface
IC	Intermediate cross-connect
IEC	International Electrotechnical Commission

IEE	Institution of Electrical Engineers
IEEE	Institute of Electrical and Electronic Engineers
InGaAs	Indium-gallium-arsenide
InGaAsP	Indium-gallium-arsenide–phosphate
IP	Internet protocol
ISO	International Organization for Standardization
LAN	Local area network
LED	Light-emitting diode
MC	Main cross-connect
MCVD	Modified chemical vapor deposition
MDF	Main distribution frame
NA	Numerical aperture
NEC	National Electronic Code
NIST	National Institute of Standards and Technology
nm	Nanometer
OCWR	Optical continuous wave reflectometer
OFC	Optical fiber conductive
OFCP	Optical fiber conductive plenum-rated
OFCR	Optical fiber conductive riser-rated
OFN	Optical fiber nonconductive
OFNP	Optical fiber nonconductive plenum-rated
OFNR	Optical fiber nonconductive riser-rated
OGW	Optical ground wire
OH+	Hydroxide radical
OLTS	Optical loss test set
OPGW	Optical power ground wire
ORL	Optical return loss
OSP	Outside plant
OTDR	Optical time domain reflectometer
OVD	Outside vapor deposition
PC	Physical contact
PCS	Personal communications systems
PVC	Polyvinyl chloride
RBOC	Regional Bell Operating Companies
RFI	Radio frequency interference
RFP	Request for proposal
RT	Room temperature
SDH	Synchronous Digital Hierarchy
SFF	Small form factor
SONET	Synchronous Optical Network
STL	Standard Telecommunications Laboratories

STP	Shielded twisted pair
TA	Technical advisory
TC	Telecom closet
telco	Telephone company
TIA	Telecommunications Industry Association
UL	Underwriters Laboratories
UTP	Unshielded twisted pair
VAD	Vapor axial deposition
WDM	Wavelength division multiplexing

INDEX